①国家社会科学基金年度项目：畲汉民族服饰艺术交流交融史证研究（22BMZ064）
②浙江省社科联人文社科出版资助项目（19YB58）
③福建省科技厅引导性项目：福建地域特色服饰文化与文化遗产保护再利用数字化技术研究与应用示范（2020H0046）
④福建省服装创意设计中心（闽江学院）开放基金资助项目（MJ2019Z001）

有凤来仪

——美丽的畲族凤凰装

The Phoenix Extends Its Grace：
Beautiful Phoenix Costumes of the She Ethnic Group

闫晶 等 / 著

中国纺织出版社有限公司

内 容 提 要

本书是介绍畲族服饰的普及读物，首先简要说明畲族概况；其次介绍了闽浙黔三地关于凤凰装的不同传说，旨在提供对畲族服饰进行多角度解读的途径；再次对典型畲族服饰进行呈现，并将承载于其上的传统手工艺进行说明；最后梳理各类型畲族服饰之间的变化规律和联系。

本书中英文对照，语言生动有趣，图文并茂，适合作为国内外对畲族服饰文化感兴趣的大众读者的通识读物。

图书在版编目（CIP）数据

有凤来仪：美丽的畲族凤凰装 / 闫晶等著. -- 北京：中国纺织出版社有限公司，2022.12

ISBN 978-7-5229-0222-7

Ⅰ.①有… Ⅱ.①闫… Ⅲ.①畲族—民族服饰—服饰文化—研究—中国 Ⅳ.①TS941.742.883

中国版本图书馆 CIP 数据核字（2022）第 253798 号

YOUFENG LAIYI: MEILIDE SHEZU FENGHUANGZHUANG

责任编辑：亢莹莹 魏 萌　　特约编辑：苗 苗

责任校对：寇晨晨　　责任印制：王艳丽

中国纺织出版社有限公司出版发行

地址：北京市朝阳区百子湾东里 A407 号楼 邮政编码：100124

销售电话：010—67004422 传真：010—87155801

http: //www.c-textilep. com

中国纺织出版社天猫旗舰店

官方微博 http://weibo.com/2119887771

北京华联印刷有限公司印刷 各地新华书店经销

2022 年 12 月第 1 版第 1 次印刷

开本：889 × 1194 1/20 印张：8

字数：154 千字 定价：69.80 元

前　言
Preface

畲族服饰，作为畲族最具辨识度的外在文化符号，早在19世纪就被国内外民族学、人类学学者关注。近年更随着旅游产业的兴起和对“非遗”文化、民间文化的重视而不断出现在大众视野中。然而由于畲族分散在闽浙赣黔等地区，各地畲族服饰各有特色，故在实际应用和传播中常出现各地域畲族服饰混淆的现象。本书据此立意，从空间的维度对覆盖畲族人口超过90%的闽浙赣黔等地的9类较为典型的畲族凤凰装款式进行呈现，以图为主，文为辅，用凝练直观的形式方便读者对众多畲服款式进行区分和记忆。同时兼顾时间的维度，将在畲族历史上留下重要痕迹的部分形制以及在当代发展中显示出令人瞩目的变化类型作为副线进行简要介绍。以期用简明的形式，通过时空结合的方式共同勾勒出一个较为整体的畲族服

The She costumes, as the most visible cultural symbol of the She people, have been paid heed to by anthropologists at home and abroad since the 19th century. In recent years, with the rise of the tourism industry and the emphasis on non-heritage culture and folk culture, the She costumes have been gaining increasing attention. Since the She ethnic group live in different provinces, i.e. Fujian, Zhejiang, Jiangxi and Guizhou, the She costumes in each region have their own characteristics, resulting in confusion in publicity. Based on this, the book presents nine typical She costumes (phoenix costumes), covering more than 90% of the She population in mentioned four provinces. Combined with pictures and words, it tries to help readers to tell and memorize different costumes in a simpler and more direct way. In this process, certain types of costumes with historic meaning and their modern variants are introduced. That being said, a more holistic picture of the She costumes is outlined in the most concise form,

饰全貌。

同时，本书结合“民族文化进校园、进课堂”的举措，针对畲族文化的主要传承人——畲族青少年，在编写中充分考虑了其作为畲族民族中小学的校本课程配套教材的实用性和适读性，设置了课后习题和扩展阅读。建议课程周期安排在16课时左右，第一、第二章1课时，第三章9课时，第四章5课时，第五章1课时。作为小学教材时，可将第五章和“扩展阅读”部分作为选修内容，最后一个课时作为总结复习。

本书在编写上注重国际性、审美性和通俗性。国际性：全书以中英文对照方式呈现，包括所有插图名及注释；在知识产权方面，以脚注形式注明书中出现的所有图片的出处。审美性：本书围绕畲族服饰体系中具有代表性和美感的畲族女性盛装展开，全书共有相关图片近300幅；为给读者留下强烈的视觉印象，本书以原创插画作为开篇对九个具有代表性的畲族服装款式进行介绍，并辅以穿着相应畲服的Q版“中国畲娃”，尽可能将畲族服饰的美以艺术表现手法进行提炼强调。通俗性：用容易理解和记忆的服饰特征为各类型凤凰装取名，替代以往延续人类学惯例的地域取名法；以畲家

considering the spheres of time and space.

The She adolescence are the main inheritors of the She culture. To better meet their demand, this book, while echoing the initiative of "ethnic culture into the school and classroom", includes after-class exercises and further reading to increase its practicality and readability as supporting material for the school-based curriculum of primary and secondary schools of the She ethnic group. The proposed curriculum cycle is about 16 durations, one duration for Chapter 1 & 2, nine durations for Chapter 3, five durations for Chapter 4 and one duration for Chapter 5. For pupils, Chapter 5 and the "Further reading" are optional, and the last duration is for review.

The book features internationalization, aestheticism and popularity. Internationalization: The book is presented in Chinese and English, including pictures and annotations; in terms of intellectual property rights, all pictures appearing in the book are footnoted with their sources. Aestheticism: The book focuses on women's costumes of the She, the most representative and aesthetically pleasing dresses in the She costumes system. Nearly 300 pictures are included. To impress readers with visual beauty, this book takes original Chinese paintings as the opening pages for the nine representative styles of *She* costumes, supplemented by the mascot of "She girl" in the corresponding She costumes. In this way, the beauty of the She costumes is refined and

小女孩“小畲凤”的口吻展开，增加亲切感的同时通过口语化的表达营造生动有趣的风格。

本书主编为闫晶，副主编为陈栩，书中插画均由林俊文创作，孙磊负责统稿。

书中所用的卡通形象（除耳巾式外）均为林章明先生设计的中国畲娃形象，特此声明并致以感谢！

本书第一章和第五章由广西大学英语笔译专业研究生钟桂梅女士翻译，并由钟女士审校全文英语。成书过程中还得到身为畲家人的钟女士及其家人的诸多建议。在此致以诚挚感谢！

同时，本书也得到印度友人思卡特（Srikant）的英语校对协助，在此一并感谢！

最后，愿所有阅读本书的读者都能通过它领略到畲族服饰的异彩纷呈！

编者

2022年8月31日

emphasized. Popularity: All phoenix costumes are named as per the characteristics for better understanding and memorization. In the past, anthropologists named by virtue of the regional naming method. But for this book, the narrative is unfolded in the oral style of the She girl, "little phenix", so readers will feel more interested and amiable.

Yan Jing is the editor-in-chief and Chen Xu associate editor. All paintings in this book are drawn by Lin Junwen. Sun Lei is the compiling editor.

The cartoon images used in this book (except the ear-shaped scarf style) are all from Mr. Lin Zhangming's design of the "Chinese She girls". Hereby we express our appreciation.

The first and fifth chapters of this book were translated by Ms. Zhong Guimei, a She lady and graduate student majoring in English translation at Guangxi University. She also helped with the proofreading of the English parts. Ms. Zhong and her family shared proposals to complete the book. Here we would like to express our sincere thanks.

The English parts were proofread by Mr. Srikant. Many thanks for the assistance from our Indian friend.

We hope our readers will enjoy the colorfulness of the She costumes!

Authors

August 31, 2022

Contents

第一章
Chapter
1

畲族概述

An overview of the She

在中国东南部的广大山区，生活着一群勤劳、勇敢、善良、忠厚的人们。一千多年来，他们过着刀耕火种的游耕生活，因此被称为“畲[1]族”。

早在公元7世纪初，即隋唐之际，畲族人民就已在闽、粤、赣三省交界地区生息、繁衍、劳动。汉文史书通常把唐宋时期在这一带活动的少数民族泛称为“蛮”[2]“蛮僚”[3]“峒蛮”[4]或“峒僚”[5]。直到13世纪中期，即南

In the vast mountainous areas of southeast China, there lived a group of industrious, brave, kind and honest people. Known as the She people[1], they have led a slash-and-burn lifestyle for more than a thousand years.

As early as the 7th century (about the turn of the Sui and Tang dynasties), the She had lived and worked in the border areas of Fujian, Guangdong and Jiangxi provinces. Generally speaking, historical documents written in Chinese termed minorities in this area from that time span as "Man[2]" (barbarian), "Manliao[3]" , "Dongman[4]" (barbarian living in the caves) or "Dongliao[5]" (barbarian from the mountains). It was the middle of the 13th century (the final years of the Southern Song Dynasty) that witnessed the appearance of "畲民" and "畬民" (literal pronunciation in Chinese for both is "She min", "畬" is the

[1] 畲，shē，意为“刀耕火种”。
She means "slash and burn". The pronunciation of She is close to [Shə].

[2]《资治通鉴》，卷二五九，《唐纪》，卷七五。
History As A Mirror, Vol. 259, *Chapter of Tang*, Vol. 75.

[3]（清光绪）《漳州府志》，卷四二，《艺文》二；（清嘉庆）《云霄厅志》（民国版），卷一一，《宦绩》；《宋史》，卷四一，《理宗》，卷一。
Local History of Zhangzhou Fu, Vol. 42 (*Art and Literature*, Vol. 2), *Yunxiao Ting Zhi*, Vol. 11 (*Officials'Performance*), *The History of Song*, Vol. 41 (*Emperor Lizong of Song*, Vol. 1).

[4]（清嘉庆）《云霄厅志》（民国版），卷一一，《宦绩》。
Yunxiao Ting Zhi, Vol. 11 (*Officials' Performance*).

[5]《宋史》，卷四一九，《许应龙传》。
The History of Song, Vol. 419 (*Biography of Xu Yinglong*).

宋末年，史书上才开始出现“畲民”和“畬民”两词并用的称呼。

“畲”字来历甚古，《诗经》《易经》当中就已有之。其字义为开荒辟地、刀耕火种。而“畲”字被用作民族的族称，始于南宋末年。元代以来，“畲民”逐渐被作为畲族的专有名称，普遍出现在汉文史籍上。

1956年国务院正式公布确定这个民族的族称为畲族，确认畲族是一个具有自己特点的单一的少数民族。贵州省的畲族是在1996年被确认的，之前被称为“东家”。

畲族自称“山哈”或“山达”，意为“山里人”或“居住山里的客人”。这个名称在史书中未有记载，但在畲族民间普遍流传。

畲族同汉族及我国许多少数民族一样，是经过长期历史发展而形成的，有着悠久的历史和文化。

畲族女子的传统服饰色彩斑斓、独具特色，被称为“凤凰装”。

ancient word for "畲", "民" is translated into "min", meaning people) in historical documents simultaneously.

The word "She" has a time-honored history, which can be found in classics like *Book of Songs* and *I Ching* (*The Book of Changes*). It means to cultivate the wasteland, to slash and burn. The word was adopted as a name for an ethnic group in the final years of the Southern Song Dynasty. Since the Yuan Dynasty, "She people" was an exclusive term for the She ethnic group, which was popular in historical Chinese documents.

1956 witnessed the official naming of "She people" by the State Council when this group was accredited as a minority people with unique characteristics. However, the She people in Guizhou province, whose former name is "Dongjia" was only officially acknowledged by 1996.

The She people called themselves "Shanha" or "Shanda", meaning "people or guests from the mountains". This name was popular in folklore but not recorded in historical books.

The She ethnic group, like other minorities, was formed after a long time and featured its long-standing history and culture.

The traditional costumes of the She women are colorful and unique, which are hailed as "phoenix costumes".

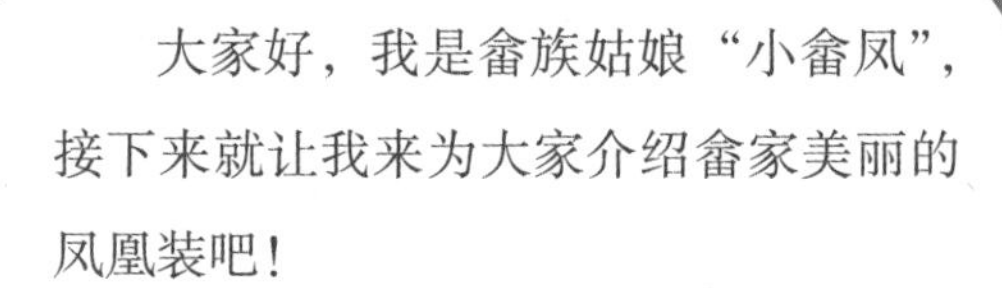

大家好，我是畲族姑娘“小畲凤”，接下来就让我来为大家介绍畲家美丽的凤凰装吧！

Hello! I am a girl of the She ethnic group. Everybody calls me "Little Phoenix". Let me introduce the beautiful She Phoenix costumes to you. Here we go!

第二章
Chapter
2

畲族凤凰装的传说

Legends of the
She Phoenix Costumes

小畲凤：为什么叫“凤凰装”？

Little Phoenix: Why "phoenix costumes"?

小畲凤：你们一定好奇为什么我们的衣服要叫“凤凰装”吧！这不仅是因为身着这些美丽服饰的畲族女子恰似传说中五彩绚丽的凤凰，还因为畲家人千百年来代代相传着不少有关凤凰和服饰的动人传说哦！

Little Phoenix: You must be wondering why our costumes are called "phoenix costumes". Is it for She women in beautiful costumes comparable to legendary phoenixes? Or is it for legends of phoenix and costumes passing down from generation to generation over past thousand years? Or maybe both? Let's check it out.

一、传说一[1]

1 Legend One[1]

据畲族史诗《高皇歌》(也叫《盘瓠王歌》)记载，畲族始祖五色神犬盘瓠生于高辛帝皇后耳中，因平番有功，金钟下变身为人后娶三公主为妻，而后定居广东转徙闽浙(图2-1)。相传盘瓠与三公主成婚后，皇后给女儿

According to the She Epic *Song of Gaohuang* (*Song of Panhu King*), the five-color supernal dog Panhu, the first ancestor of the She, was born in the ear of Emperor Gaoxin's wife. After Panhu defeated the enemy, he married the third Princess and became a human being under a golden bell. Later on, Panhu

戴上凤冠，穿上镶着珠宝的凤衣，希望她像凤凰一样给生活带来祥瑞。三公主有了女儿后，也把她打扮得像凤凰一样。从此，畲家女便沿袭了三公主的装束，着“凤凰装”，她们用红头绳扎的头髻象征着凤冠；在衣裳胸前和围裙上刺绣出各种彩色花边，并镶绣着金丝银线，象征着凤凰的颈部、腰部美丽的羽毛；后腰随风飘动的艳丽腰带，象征着凤凰的尾巴；周身悬挂着叮当作响的银器，象征着凤凰的鸣啭。

settled in Guangdong and migrated to Fujian and Zhejiang(Fig. 2-1). According to the legend, upon the wedding day, the empress dressed her daughter in a crown featuring a phoenix and a jeweled phoenix garment, wishing to bring more fortune to her beloved daughter. When the third Princess had a daughter, she dressed her up also like a phoenix. From then on, the She women followed the dressing style of the third Princess, wearing the "phoenix costumes". The head bun tied with a red cord symbolizes the phoenix crown; the colorful laces embroidered with gold and silver thread in the front of the dress and apron are symbols of the beautiful feathers on the phoenix's neck and waist; the bright belts fluttering in the wind at the back are symbols of the phoenix's tails; and silver jewelry decorated on the costumes sounds like phoenix's warble.

图2-1 祖图[1]

Fig. 2-1 Ancestral figure

[1] 图片来源：2005年3月13日摄于浙江省丽水市云和县雾溪乡江南畲族风情文化村。

Source: photo was taken in the Jiangnan She Culture Village, She Township of Wuxi, Yunhe County, Lishui, Zhejiang on Mar. 13, 2005.

二、传说二[2]

2 Legend Two[2]

浙江丽水、云和、景宁一带的畲族流传着其祖先为“凤父龙母”所诞子孙的传说。相传广东潮州附近有座凤凰山，山形像凤凰。有一天，天上飞来一只金凤凰，凤凰吃了白玛瑙，产下一颗凤凰蛋，凤凰蛋里蹦出个胖娃娃，取名为凤哥，并由百鸟抚育成人，后来朝着太阳升起的东方走去，打死大蟒，打败猛虎，历尽艰险，娶龙女为妻，繁衍的后代即为畲族子孙，从此以后，畲族人以凤凰和龙为图腾，逢年过节均举行敲锣泼水的祭祀祖先活动。

The She people in Lishui, Yunhe and Jingning of Zhejiang Province still believe they are offspring of phoenix and dragon. Phoenix Mountains in Chaozhou, Guangdong, as said, resembled a phoenix. One day, a golden phoenix flew there from heaven. After eating a white agate, the phoenix produced an egg, which was incubated and grown to be a fat baby, named Fengge (phoenix brother). Reared by all types of birds, Fengge grew up and headed east towards the sun rises. Along the journey, Fengge killed a huge python, defeated a fierce tiger, and endured all kinds of hardships and adversities. Finally, he made it and married a dragon lady, and their offspring became the ethnic group of the She. Since then, the She people had taken the figures of the phoenix and dragon as their totems, and hold ancestor worship activities on holidays every year such as gong beating and water splashing.

三、传说三[3]

3 Legend Three[3]

关于凤凰装的由来，贵州畲族《开路径》中作了这样的描述：开天辟地之后，龙、虎、雷公等为争夺天下，展开激战，龙发大水淹没低处，雷击电闪，引燃大火焚烧陆地山冈，风狂火猛，水势汹涌，百物逃无所逃，遁无所遁，霎时陷入毁灭绝境。正当危急时刻，凤凰长声高叫，振翅冲天，突出了火的重围，然后迫使龙潜于渊，雷藏于天，才又招回百物，人们重新开始了幸福祥和的生活。凤凰是百鸟之王，女性的象征，具有大智大勇的特点，是畲族人的保护神，因其对畲族有再生之德，翼护之恩，大多将凤凰图案作为女性饰物，故称女性盛装为“凤凰衣”“凤凰装”。

Regarding the origin of phoenix costumes, *The Open Path* of the Guizhou She ethnic group unfolds the story as below. At the very beginning when heaven and earth were created, Dragon, Tiger and Thunder God battled for the reign: the Dragon stirred a flood overflowing the low places, while thunder and lightning inundated the lands and hills. Facing the wildfire and water surges, nothing escaped, and all suddenly descended to ruin. In this critical moment, a phoenix cried in a long and high voice, flapped its wings into the sky, and flew out of the fire encirclement. The phoenix finally forced the Dragon to dive into the deep, and the Thunder God to hide in the sky. Only then did all things return to their happy and peaceful ways. The phoenix thus became the king of birds and the symbol of women, boasting enormous wisdom and courage and protecting and blessing the She. That explains why most She people decorate their clothes with phoenix patterns and call such clothes "phoenix costumes".

思考题 | Exercise

请大家思考畲族各地区的服饰与当地传说之间的关系。

Please try to connect the legends with the local costumes.

第三章
Chapter
3

各地凤凰装的特色

Styles of the She Phoenix Costumes

小畲凤：原来各地不一样！

Little Phoenix: It varies from place to place!

小畲凤：一千多年来，我的祖先们不畏艰难险阻，开山辟土，从原始居住地广东省潮州市凤凰山四散迁徙到福建、浙江、江西、安徽等省，有的还到达贵州和四川（图3-1）。在这个过程中，他们的服饰也因为受到当地文化的影响而呈现出丰富多彩的变化。各地畲族凤凰装从而异彩纷呈，各有千秋。这里选取了其中比较有代表性的九种款式来介绍。

由于各地凤凰装丰富多彩，从衣裤到首饰都不一样，为了便于大家区分和记忆，以下用凤凰装里最引人注目，也最具有特色的头冠[1]

Little Phoenix: For more than a thousand years, our ancestors encountered various dangers and difficulties to reclaim the wasteland. They left Phoenix Mountains in Chaozhou, Guangdong, and migrated to Fujian, Zhejiang, Jiangxi, Anhui, and even reached as far as Guizhou and Sichuan provinces (Fig. 3-1). During this process, our costumes also presented a rich variety of new trends under the influence of local cultures. Therefore, the She phoenix costumes vary from place to place; each has its own merits. Here are nine representative styles.

As phoenix costumes differentiate from each others ranging from garments to jewelry, we use crown, the most eye-catching and distinctive part of the phoenix costumes, to name the costumes[1].

---

[1] 人类学家凌纯声先生早在1947年就对各地畲族妇女头饰造型进行研究，认为畲族妇女头饰为唯一“在外表观察上可以区别的”畲族特征。
As early as 1947, Mr. Ling Chunsheng studied the headwear of She women in different places and believed that headwear was the only characteristic that differentiated the people from others.

给各式凤凰装命名。

最早对凤凰装进行分类和命名的是人类学家们。1947年凌纯声先生在《畲民图腾文化的研究》一文中对畲族女性头饰进行了分类和命名，分为丽水道士畈式——简称丽水式、景宁敕木山式——简称景宁式和福州罗岗式——简称福州式。20世纪60年代，蒋炳钊先生进一步将福建畲族头冠分为三种类型。1985年，潘宏立先生在硕士论文《福建畲族服饰研究》中通过福建各地畲族女性整体装束的特征差异，将之区分为罗源式、福安式、霞浦式、福鼎式、顺昌式、光泽式、漳平式七种类型。1999年，《浙江省少数民族志》将浙江畲族妇女头饰分为景宁式、丽水式、平阳式、泰顺式。他们多以畲族所聚居的县（市）名来命名凤凰装，但对于不太熟悉浙闽地理的读者来说，可能过于抽象而难以理解和记忆，因此本书以服饰特征对其命名。

In this connection, it is easier for you to tell and know them.

Anthropologists were the first to classify and name Phoenix costumes. In 1947, Mr. Ling Chunsheng classified and named the She female headdress in the paper *A Study on She Totem Culture*, among which several styles were introduced, i.e. Lishui Daoshibe style (Lishui style), Jingning Chimushan style (Jingning style), and Fuzhou Luogang style (Fuzhou style). In the 1960s, Mr. Jiang Bingzhao divided the She crowns in Fujian into three types. In 1985, Mr. Pan Hongli studied the differences in characteristics and proposed seven types of She women costumes in Fujian in his thesis Research on She Clothing in Fujian, that is, Luoyuan style, Fu'an style, Xiapu style, Fuding style, Shunchang style, Guangze style and Zhangping style. In 1999, *The Annals of Ethnic Minorities in Zhejiang Province* divided She women's headwear in Zhejiang into Jingning style, Lishui style, Pingyang style and Taishun style. All the above styles are named after the county (city) where the She people live. As readers who are unfamiliar with the geography of Zhejiang and Fujian might find it difficult to understand, not to mention to memorize, this book names the styles after the features of the costumes.

图3-1　畲族及其代表性服饰分布示意图[1]

[1] 红色地名为目前畲族相对聚集的县（区）。

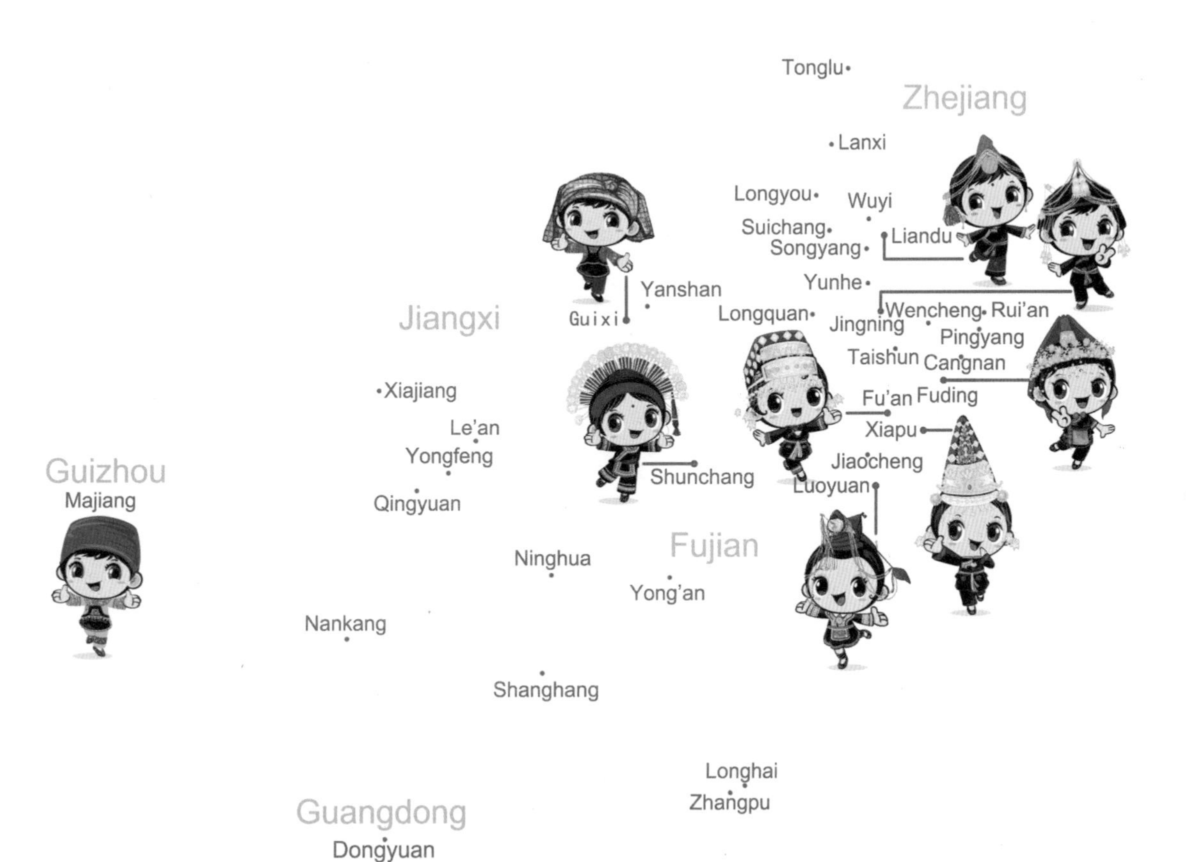

Fig. 3-1 Distribution of the She and their representative costumes[1]

[1] Places in red are the counties (districts) where current She ethnic group live in individual concentrated communities.

一、雄冠式

1 Male Crown Style

位于浙江省南部的景宁县是一座依山傍水、风景秀丽的小城，也是我国唯一的畲族自治县（图3-2）。在那里，畲家女子用高高的银饰和雪白的珠链装点自己。这神气的头饰就像昂首挺胸的雄凤凰，所以我们骄傲地称它为“雄冠”（图3-3~图3-8）。

雄冠式凤凰装典型服装为“花边衫”（图3-9）。“花边衫”为长度及臀的大襟上衣和阔脚直筒长裤，它的特色是上衣襟边依次间隔镶有红、蓝色贴布条，袖口有一条或蓝或白的细贴边。

Jingning County, the only She autonomous county located in the south of Zhejiang Province, is a small county with beautiful scenery(Fig. 3-2). The She women there wear tall silver ornaments and white beads. The perky headdress is like a male phoenix with its head held high, so we proudly call it "the male crown" (Fig. 3-3~Fig. 3-8).

The phoenix costume in male crown style is featured by a lace garment (Fig. 3-9), which includes a hip-length garment and a pair of wide-legged straight trousers. The garment has red and blue strips at intervals along the hem of the top, and a blue or white thin hem at the cuffs.

(a)

(b)

图3-2 景宁畲乡风光[1]

Fig. 3-2 Sceneries of She Villages in Jingning, Zhejiang

[1] 图片来源：(a)：笔者于2004年3月9日摄于景宁县城鹤溪镇东弄村，(b)：笔者于2005年3月11日摄于景宁县渤海镇郑坑村。
Source: the pictures were taken in Dongnong Village, Hexi Town, Jingning County on March 9, 2004, and Zhengkeng Village, Bohai Town, Jingning County on March 11, 2005, respectively.

图3-3 穿着雄冠式畲族服饰的“中国畲娃”卡通形象

Fig. 3-3 Cartoon image of "Chinese She Girls" in male crown style

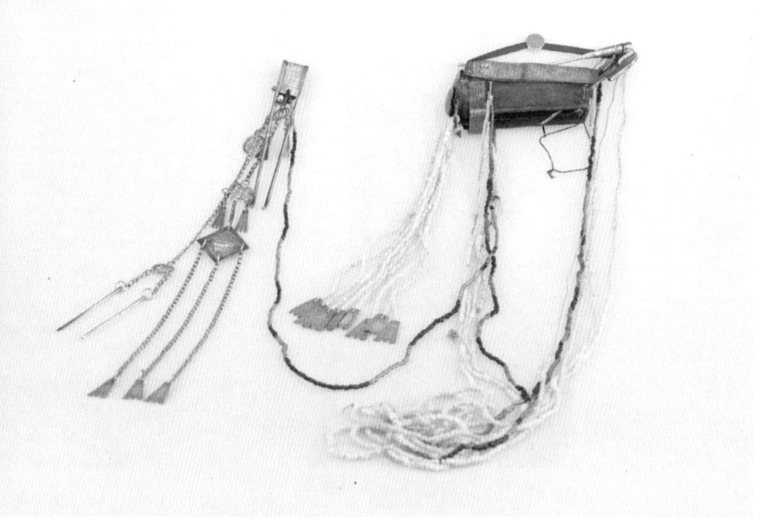

图3–4 雄冠[1]

Fig. 3-4 Male crown

图3–5 雄冠前视图

Fig. 3-5 The front view of the male crown

图3–6 雄冠主体

Fig. 3-6 The main part of the male crown

---

[1] 图片来源：图3–4～图3–8由景宁中国畲族博物馆提供。

Source: Fig. 3-4 to Fig. 3-8 provided by She Museum of China in Jingning.

图3-7　雄冠主体后部

Fig. 3-7　The back of the male crown

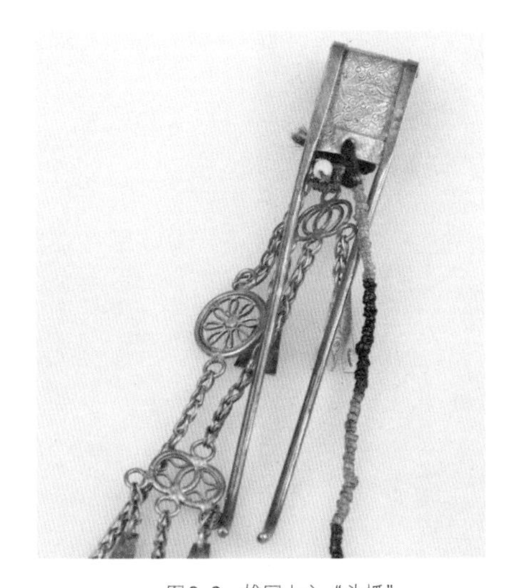

图3-8　雄冠上之“头抓”

Fig. 3-8　The "Touzhua" of the male crown

图3-9　雄冠式凤凰装[1]

Fig. 3-9　Phoenix costume in male crown style

---

[1] 图片来源：2005年3月12日摄于浙江省景宁畲族自治县渤海镇郑坑村。

Source: the photos were taken at Zhengkeng village, Bohai Town, She Autonomous County, Jingning, Zhejiang Province on Mar. 12, 2005.

扩展阅读
Further Information

历史上的景宁畲族服饰
Jingning She Costumes in History

1929年夏天，德国学者史图博和中国学者李化民走访景宁，撰写了《浙江景宁县敕木山畲民调查记》。根据他的记载，当时“畲族妇女们普遍穿着老式剪裁的无领上衣，领口和袖口上镶阔边”（图3–10）。

In the summer of 1929, the German scholar Stübel H and the Chinese scholar Li Huamin visited Jingning and completed *Die Hsia-min vom Tse-mu-schan* (*The Investigation of She Ethnic People in Chimu Mountain*). According to Stübel, "it is common for She women to wear collarless garment tailored in traditional fashion, with broad rims at the collar and cuffs" (Fig. 3-10).

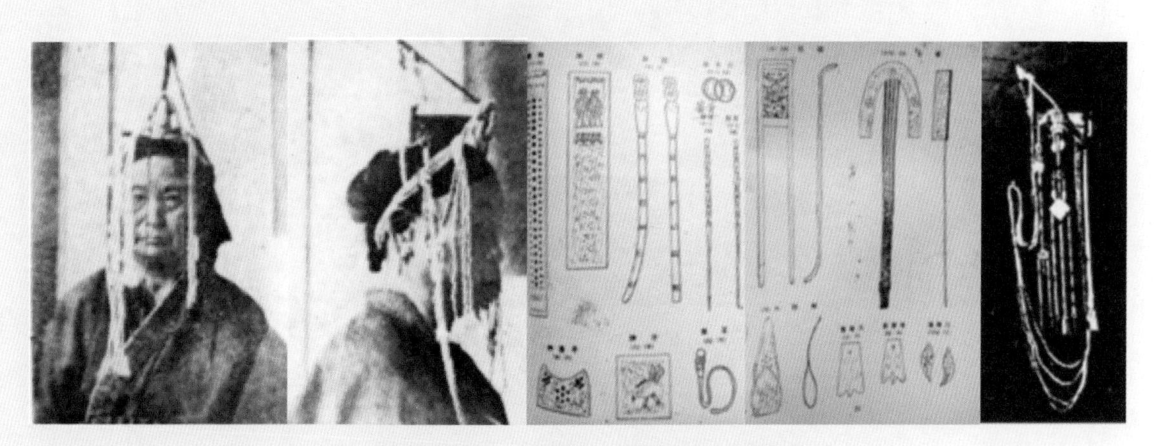

图3–10 《敕木山畲民调查记》中景宁式畲族头饰插图
Fig. 3-10 She headwear in Jingning in *Die Hsia-min vom Tse-mu-schan*

发展中的雄冠式凤凰装（图3-11、图3-12）

The Development of the Male Crown Style (Fig. 3-11, Fig. 3-12)

图3-11 畲族博物馆藏现代设计版景宁凤凰装

Fig. 3-11 The modern version of the Jingning phoenix costumes housed in the She Museum of China in Jingning

图3-12 第4届中国（浙江）畲族服饰设计大赛获奖作品[1]

Fig. 3-12 The 4th China (Zhejiang) She ethnic group Fashion Design Competition winning costumes

---

[1] 图片来源：2018年4月18日，第四届中国（浙江）畲族服饰设计展演。
Source: The 4th China (Zhejiang) She ethnic group Fashion Design Exhibition, 2018-4-18.

练习题 | Exercise

请为左图的畲族服饰填上你喜欢的色彩吧！

Please color in picture for the She costumes.

二、雌冠式

2 Female Crown Style

在浙江省丽水市，除景宁县以外，聚居在云和县、松阳县、遂昌县、缙云县、青田县、龙泉市和莲都区等地的畲民几乎均着类似形制的凤凰装（图3–13）。

In Lishui City, Zhejiang, the She people living in Yunhe County, Songyang County, Suichang County, Jinyun County, Qingtian County, Longquan City and Liandu District wear similar phoenix costumes, which are different from that of Jingning (Fig. 3-13).

图3–13 丽水畲乡风光[1]

Fig. 3-13 Scenery of the She Village in Lishui, Zhejiang

[1] 图片来源：2005年3月13日摄于丽水地区云和县雾溪畲族乡。
Source: photo was taken in the She Township of Wuxi, Yunhe County, Lishui District on March 13, 2005.

比起耀眼夺目的雄冠式凤凰装，丽水其他地区的畲族女子的头饰相对简朴，犹如雌凤凰，因此被称为“雌冠式”（图3-14~图3-16）。

Compared with the exquisite male crown style in Jingning, the headdresses of the She women in those areas are relatively simpler. Similar to female phoenixes, this type of costume is called the female crown style (Fig. 3-14~Fig. 3-16).

（a） （b）

图3-14 雌冠式头饰[1]

Fig. 3-14 The female crown

[1] 图片来源：（a）：金成禧.畲族传统手工织品——彩带[J].中国纺织大学学报，1999（25）：100；（b）：2010年4月27日翻拍于厦门大学人类学博物馆，原图摄于浙江省丽水市老竹镇。

Source: (a): Kim Sung Hee. Traditional Handmade Fabric of She ethnic group — Colorful Ribbon [J]. Journal of China Textile University, 1999 (25): 100; (b): taken at the Museum of Anthropology of Xiamen University on April 27, 2010. The original photo was taken in Laozhu Town, Lishui City, Zhejiang Province.

图3–15 穿着雄冠式畲族服饰的“中国畲娃”卡通形象

Fig. 3-15 Cartoon image of "Chinese She Girls" in male crown style

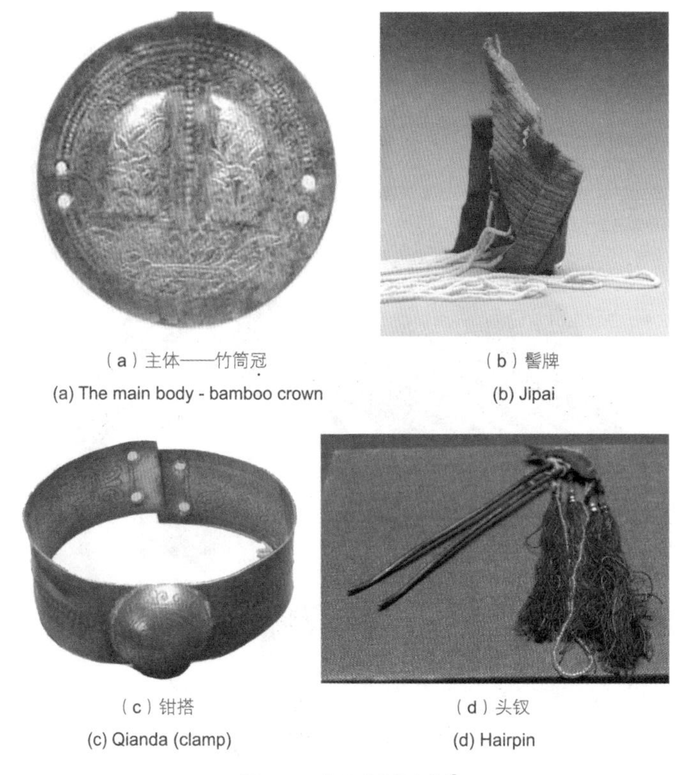

（a）主体——竹筒冠
(a) The main body - bamboo crown

（b）髻牌
(b) Jipai

（c）钳搭
(c) Qianda (clamp)

（d）头钗
(d) Hairpin

图3-16　雌冠式头饰细部[1]
Fig. 3-16　Details of the female crown

与雌冠搭配的是如图3-17所示花边衫。立领、连身袖、右衽、收腰、侧边开衩。面料为深色棉布，领口及襟边镶浅蓝色绲边，大襟边贴镶彩色织带。

The female crown is matched with a lace garment as shown in Fig. 3-17. It has an upright collar, kimono sleeves, right lapel, nipped waist and side slits. The fabric is made of dark cotton, with light blue piping at the neckline and front edge, and colorful ribbons at the front edge.

---

[1] 图片来源：（a）：征集于丽水碧湖公社高溪大队，1959年收藏于浙江省博物馆；（b、c）：征集于丽水地区英雄赤坑村，1959年收藏于浙江省博物馆；（d）：征集于丽水地区，1959年收藏于浙江省博物馆。
Source: (a): collected from the Gaoxi Brigade of Bihu Commune in Lishui and housed in Zhejiang Provincial Museum in 1959; (b,c): collected from Yingxiong Chi Keng Village, Lishui Gty, and housed in Zhejiang Provincial Museum in 1959; (d): collected in Lishui area and housed in Zhejiang Provincial Museum in 1959.

（a）

（b）

图3–17 雌冠式畲族盛装[1]

Fig. 3-17 She formal costume in female crown style

[1] 图片来源：（a）：金成禧.畲族传统手工织品——彩带[J].中国纺织大学学报，1999（25）：100；（b）：2005年3月13日摄于浙江省丽水市云和县雾溪乡江南畲族风情文化村。

Source: (a): Kim Sung Hee. Traditional Handmade Fabric of She ethnic group — Colorful Ribbon [J]. Journal of China Textile University, 1999 (25): 100; (b): taken in Jiangnan She Nationality Culture Village, Wuxi Town, Yunhe County, Lishui City, Zhejiang on Mar. 13, 2005.

扩展阅读
Further Information

历史上的丽水畲族服饰
Lishui She Costumes in History

1934年5月，凌纯声、芮逸夫、勇士衡等人到浙江旧处州府所属丽水、景宁、云和、遂昌、松阳、龙泉、宣平（今属武义县）等地考察畲族生活状况和社会生活（图3-18）。

Mr. Ling Chunsheng, Mr. Rui Yifu, and Mr. Yong Shiheng field studied Lishui, Jingning, Yunhe, Suichang, Songyang, Longquan, and Xuanping (now Wuyi County) in May of 1934 to peep the living conditions of the She people there (Fig. 3-18).

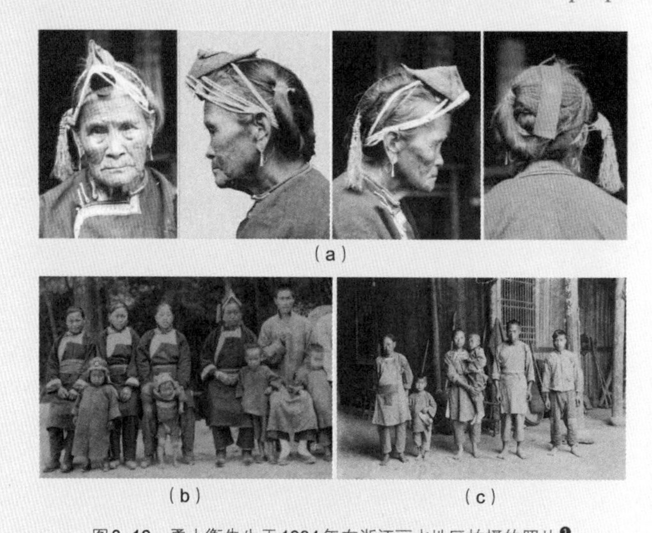

图3-18　勇士衡先生于1934年在浙江丽水地区拍摄的照片[1]

Fig. 3-18　The photos taken by Mr. Yong Shiheng in Lishui, Zhejiang in 1934

---

[1] 图片来源：(a)：1934年勇士衡摄，原图存于中央研究院历史语言研究所；(b)：为丽水山根沃门下蓝姓家族；(c)：为丽水西山畔雷水源家族。蓝、雷、钟、盘为畲族四大姓。

Source: (a): Mr. Yong Shiheng photographed in 1934, the original picture is kept in the Institute of History and Philology, Academia Sinica; (b): represents the Lan family in Lishui; (c): represents the Lei family in Lishui. Lan, Lei, Zhong and Pan are the four main surnames of the She.

20世纪70年代后期，雌冠式头冠逐渐简化。时至今日，现代雌冠式头饰的造型已演变为一个黑色头箍，头箍中间竖起一个红色布包小三角，头箍上再缀以珠串装饰。所有装饰均固定于黑箍上，因此佩戴时非常方便，只需将黑箍套在头上即可（图3-19、图3-20）。

Since the 1970s, the female crown has gradually been simplified. Today, the modern female crown has evolved into a black headband, with a small triangle of red cloth bag erected in the middle, and then decorated with beads. All the decorations are fixed to the black headband, making it easy to wear (Fig. 3-19, Fig. 3-20).

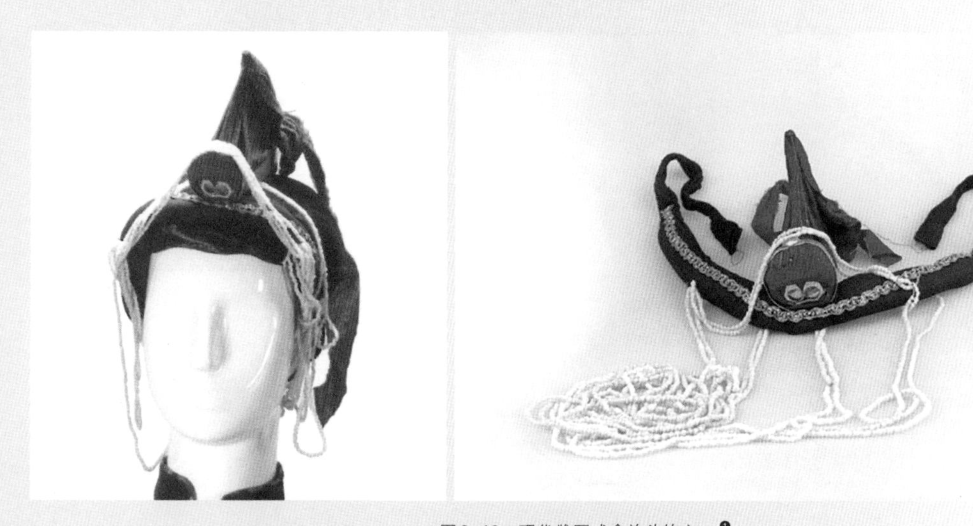

图3-19 现代雌冠式畲族头饰之一[1]

Fig. 3-19 The first modern She headwear in female crown style

[1] 图片来源：由浙江省博物馆提供，1977~2000年征集于浙江省丽水市云和县。

Source: Provided by Zhejiang Provincial Museum; collected from Yunhe County, Lishui City, Zhejiang from 1977 to 2000.

图3–20　现代雌冠式畲族头饰之二[1]

Fig. 3-20　The second modern She headwear in female crown style

❶ 图片来源：2005年3月13日摄于浙江省丽水市云和县雾溪乡江南畲族风情文化村。

Source: taken in the south of the She Nationality Culture Village, Wuxi Town, Yunhe County, Lishui City, Zhejiang on March 13, 2005.

练习题 | Exercise

请为左图畲族服饰填上你喜欢的色彩吧！

Please color in picture for the She costumes.

三、凤翎式

3 Phoenix Feather Style

从丽水出发，向东南方向继续寻访，会发现在浙江温州南部（图3–21）和福建福鼎地区的畲家凤凰装非常相似。由于这种服式在头冠脑后和胸前右襟处都有两条长长的飘带，与传统凤凰形象中的两根长尾翎相似，因此称为“凤翎式”（图3–22~图3–26）。

Starting from Lishui, and southeastwards, we find that the She phoenix costumes in the south of Wenzhou in Zhejiang province (Fig. 3-21) and Fuding in Fujian province are very similar. The Phoenix feather style is thus called due to its having two long streamers at the back of the crown and the right front of the chest; which is similar to the two long tail feathers of the traditional image of the phoenix (Fig. 3-22~Fig. 3-26).

图3–21 温州畲乡风光[1]

Fig. 3-21 Scenery of the She village in Wenzhou, Zhejiang

[1] 图片来源：摄于2004年6月19日，温州平阳县青街畲族乡。
Source: taken in She Township of Qingjie, Pingyang County, Wenzhou, Zhejiang on Jun. 19, 2004.

图3–22　穿着凤翎式畲族服饰的“中国畲娃”卡通形象

Fig. 3-22　Cartoon of "Chinese She Girls" in phoenix feather style

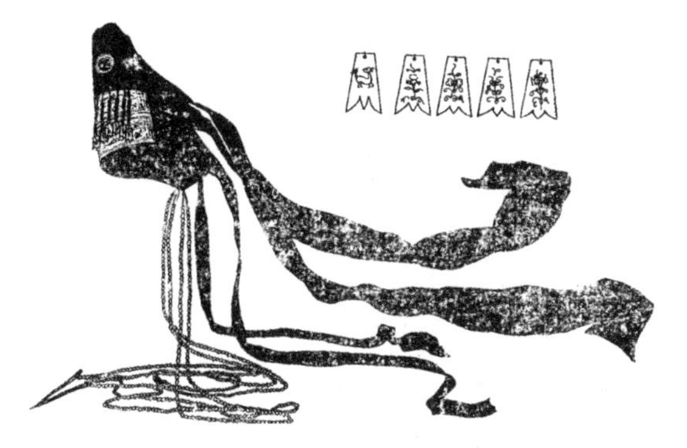

图3-23 潘宏立《福建畲族服饰研究》书中的凤翎式头饰

Fig. 3-23 The phoenix feather styled headwear in *Study of She Costumes in Fujian* written by Mr. Pan Hongli

图3-24 河南安阳殷墟妇好墓出土的双尾羽玉凤（商代）[1]

Fig. 3-24 Jaded phoenix unearthed from Fu Hao Tomb, Yin Ruins, Anyang, Henan (Shang Dynasty, BC1600-BC1046)

图3-25 浙江凤翎式畲族盛装[2]

Fig. 3-25 The formal costume in phoenix feather style in Zhejiang

---

❶ 图片来源：中国国家博物馆官网。

注：妇好为中国历史上有据可查（甲骨文）的第一位女性军事统帅，同时也是一位杰出的女政治家。

Source: Official website of National Museum of China.

Notes: It is recorded (Oracle Bone Script) that Fu Hao is the first female military commander in Chinese history, who is also an outstanding female politician.

❷ 图片来源：2004年5月4日摄于浙江省温州市苍南县凤阳畲族乡。

Source: taken in Fengyang She Township, Cangnan County, Wenzhou City, Zhejiang Province on May 4, 2004.

(a) (b) (c)

图3-26 凤翎式新娘、少女及妇女头饰[1]

Fig. 3-26 Bride, maid and women headwear in phoenix feather style

[1] 图片来源：(a、b)：由福鼎市民族与宗教事务局钟敦畅先生提供，摄于2007年太姥山旅游文化节；(c)：2010年由汤瑛女士提供。
Source: (a,b): provided by Mr. Chung Dunchang of the Administration of Ethnic and Religious Affairs of Fuding County. taken at Taimu Mountain Tourism and Culture Festival in 2007. (c): courtesy of Ms. Tang Ying in 2010.

凤翎式凤凰装的主要特征为复式领结构、“杨梅花”、前襟刺绣、飘带和假袖口（图3–27）。

Phoenix feather style features a duplex collar, myrica rubra shaped decoration, embroidery on the front, ribbons and false cuffs (Fig. 3-27).

图3–27 20世纪60年代福建凤翎式畲族女服❶
Fig. 3-27 She costume in phoenix feather style in Fujian in the 1960s

复式领结构是指衣领分大领和小领。领面颜色多选用水红、水绿。领子上的刺绣，常用图案有牡丹、莲花等花卉（图3–28）。

The duplex collar structure means that the collar is divided into a big collar and a small one. Most collars take cerise and light green as the main colors. The patterns of embroidery on the collar feature peony and lotus (Fig. 3-28).

❶ 图片来源：2010年4月23日摄于福建省福鼎市民族与宗教事务局。
Source: taken in Administration of Ethnic and Religious Affairs of Fuding City, Fujian Province on April 23, 2010.

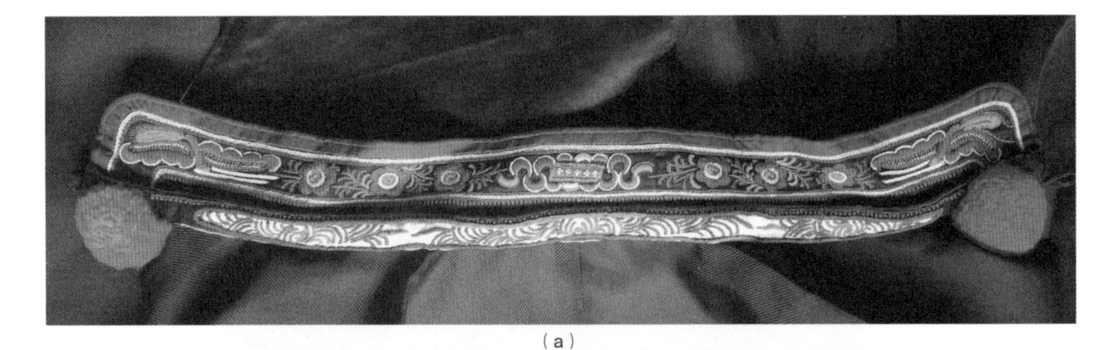

(a)

(b)

图3–28 凤翎式复式领结构及领面绣花[1]

Fig. 3-28 The duplex collar and the embroideries in phoenix feather style

---

[1] 图片来源：(a)：2010年4月23日摄于福建省福鼎市民族与宗教事务局；(b)：2004年5月4日摄于浙江省温州市苍南县凤阳畲族乡。

Source: (a): taken in the Administration of Ethnic and Religious Affairs of Fuding City, Fujian Province on Apr. 23, 2010. (b): taken in Fengyang She Township, Cangnan County, Wenzhou City, Zhejiang Province on May 4, 2004.

盛装领口装饰两颗直径约2厘米的红绿相间的绒球，球底托十几片布叶子，球心镶各色料珠，有的还饰以小银片，俗称“杨梅花”（图3-29）。

The neckline is decorated with two red and green padded balls with a diameter of about 2 centimeters. The bottom of the ball supports a dozen pieces of cloth leaves. The center of the ball is inlaid with colorful beads, and some are decorated with small silver pieces, commonly known as myrica rubra (Fig. 3-29).

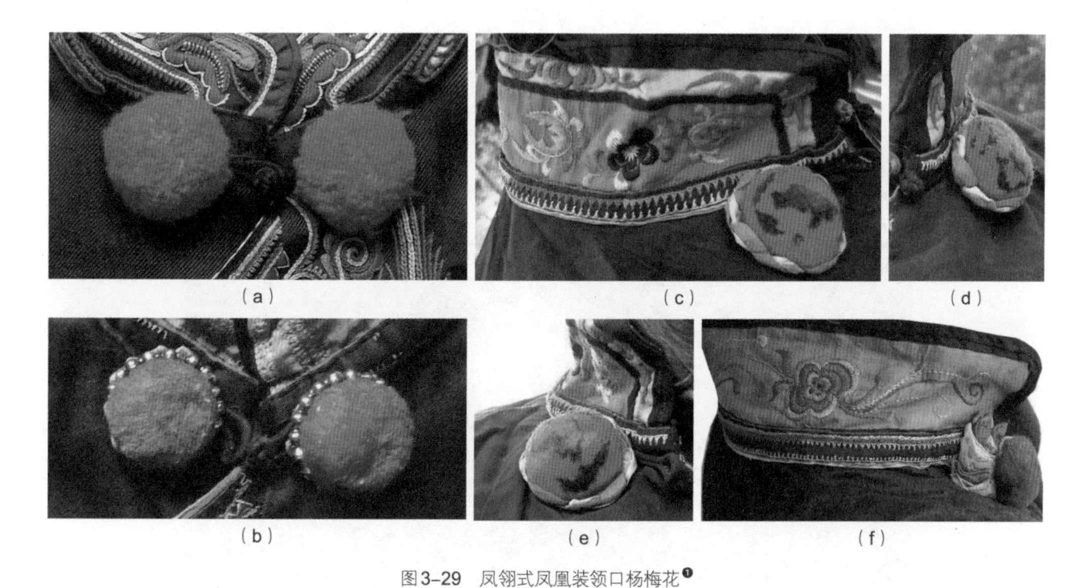
(a) (c) (d) (b) (e) (f)

图3-29 凤翎式凤凰装领口杨梅花[1]

Fig. 3-29 The myrica rubra on the collar in phoenix feather style

上衣右前襟绣有大面积色彩鲜艳的钩形适合纹样，刺绣题材灵活多变，人物及动植物造型生动活泼（图3-30）。

The right front of the coat has a large area of colorful hook-shaped embroidery. The themes are colorful, with vivid and lively figures represented by various flora and fauna (Fig. 3-30).

[1] 图片来源：(a)、(b)：笔者于2010年4月23日摄于福建省福鼎市民族与宗教事务局；(c)~(f)：笔者于2004年5月4日摄于浙江省温州市苍南县凤阳畲族乡。

Source: (a),(b): taken in the Administration of Ethnic and Religious Affairs of Fuding City, Fujian Province on Apr. 23, 2010. (c) to (f): taken in Fengyang She Township, Cangnan County, Wenzhou City, Zhejiang Province on May 4, 2004.

（a） （b）

（c） （d）

图3-30 凤翎式前襟刺绣❶

Fig. 3-30 Front embroidery in phoenix feather style

❶ 图片来源：（a）、（b）：笔者于2010年4月23日摄于福建省福鼎市民族与宗教事务局；（c）、（d）：笔者于2004年5月4日摄于浙江省温州市苍南县凤阳畲族乡。

Source: (a),(b): taken in the Administration of Ethnic and Religious Affairs of Fuding City, Fujian Province on Apr. 23, 2010. (c),(d): taken on May 4, 2004 in Fengyang She Township, Cangnan County, Wenzhou City, Zhejiang Province.

右边大襟襟边腋下垂两条桃红或大红绣花飘带，长过衣裾（图3-31）。

Two embroidered ribbons of pink or crimson hang under the armpits of the right side of the garment, extending beyond the hem (Fig. 3-31).

（a）　（b）　（c）

图3-31　凤翎式飘带[1]

Fig. 3-31　The ribbons in phoenix feather style

[1] 图片来源：（a）：笔者于2010年4月23日摄于福建省福鼎市民族与宗教事务局；（b）、（c）：笔者于2004年5月4日摄于浙江省温州市苍南县凤阳畲族乡。

Source: (a): taken on April 23, 2010 in Administration of Ethnic and Religious Affairs of Fuding City, Fujian Province. (b),(c): taken in Fengyang She Township, Cangnan County, Wenzhou City, Zhejiang Province on May 4, 2004.

袖口贴边，配以红色布条，或加其他颜色的布边绲边，以模仿多层袖口。据说过去家境越殷实的人穿的衣服越多，所以畲族妇女纷纷在袖口模拟出穿有多层上衣的效果，以表富裕，渐渐便形成风俗（图3-32）。上衣两侧衣衩内缘镶红色贴条。

The cuff is hemmed with a red strip, or hems in other colors are rolled, to simulate a multi-layer hem. It is said that in the past, the richer the family was, the more clothes they wore. So, the She women tried to pretend they had wore many clothes by showing multi-layer coats on the cuffs. That was how the custom gradually formed. Inside edges of vents on both sides of the coat are rolled with red stickers (Fig. 3-32).

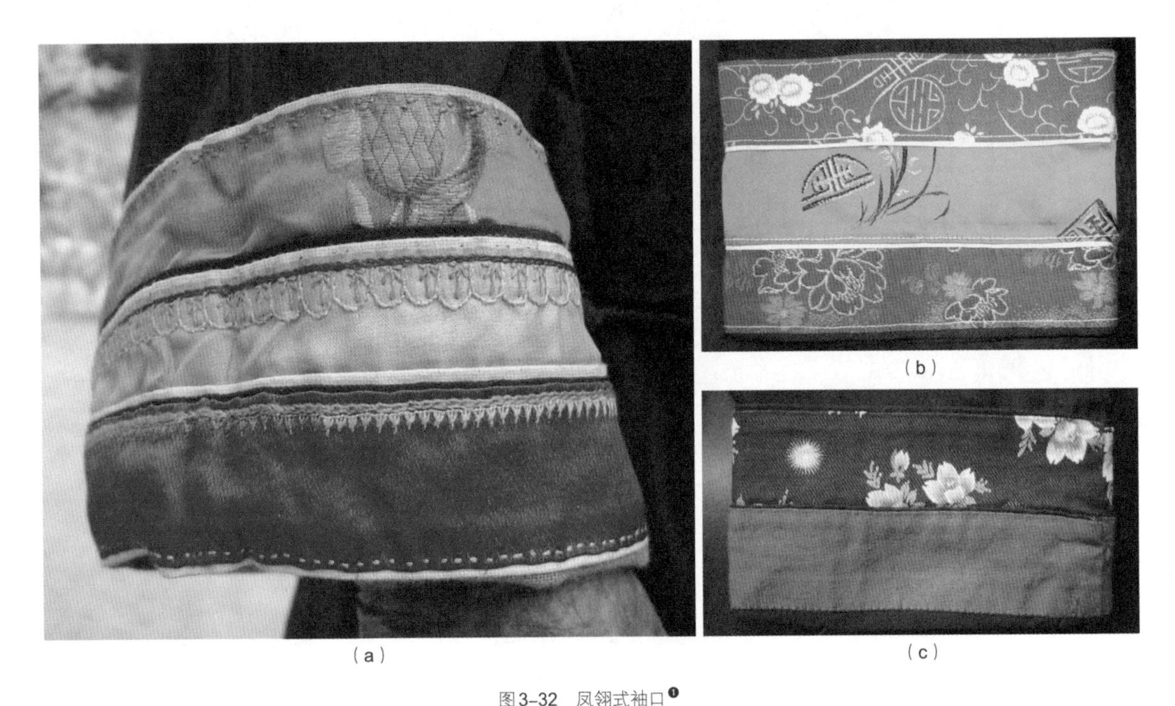

（a）（b）（c）

图3-32　凤翎式袖口[1]

Fig. 3-32　The cuffs in phoenix feather style

[1] 图片来源：（a）：笔者于2004年5月4日摄于浙江省温州市苍南县凤阳畲族乡；（b）、（c）：笔者于2010年4月23日摄于福建省福鼎市民族与宗教事务局。

Source: (a): taken in Fengyang She Township, Cangnan County, Wenzhou City, Zhejiang Province on May 4, 2004. (b),(c): taken in the Administration of Ethnic and Religious Affairs of Fuding City, Fujian Province on Apr. 23, 2010.

扩展阅读
Further Information

历史上的平阳畲族服饰

Pingyang She Costumes in History

过去，温州平阳的畲族妇女戴如图3-33所示的凤冠，还喜欢用如图3-34所示的步摇钗钏装饰自己。但是从20世纪70年代改革开放以来，平阳地区就很少能见到戴着这样头饰的人了（图3-35）。

In the past, the She women in Pingyang, Wenzhou, wore phoenix crowns as shown in Fig. 3-33, and decorated themselves with hairpins as shown in Fig. 3-34. Rare people wore them since China's Reformation and Opening-up in the 1970s (Fig. 3-35).

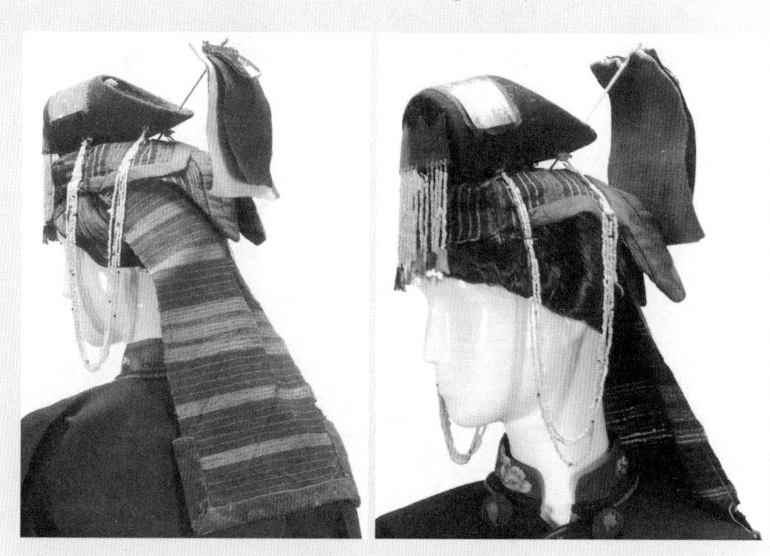

图3-33　平阳畲族凤冠[1]

Fig. 3-33　The phoenix crown of the She minority in Pingyang

[1] 图片来源：浙江省博物馆；1959年征集于平阳桥墩门公社营溪大队。

Source: Zhejiang Provincial Museum; collected from Yingxi Brigade Qiaodunmen Commune in Pingyang County in 1959.

图3-34 平阳畲族银把头[1]

Fig. 3-34 Sliver headwear of She woman in Pingyang

图3-35 《平阳畲民调查》中所载照片资料——妇女头饰（1934年）

Fig. 3-35 Women's headwear recorded in *Pingyang She Ethnic Field Study* (1934)

---

1. 图片来源：浙江省博物馆。1959年征集于平阳县桥墩公社洋尾村。
Source: Zhejiang Provincial Museum; collected from Yangwei Village Qiaodun Commune in Pingyang County in 1959.

练习题 | Exercise

请为畲族服饰填上你喜欢的色彩吧！

Please color in picture for the She costumes.

四、凤尾式

4 Phoenix Tail Style

霞浦县位于福鼎市的南边（图3-36），该地区的凤凰装因其独特的“凤尾髻”而被称为“凤尾式”（图3-37）。凤尾式除了流行于霞浦县畲族村庄，还流行于福安市东南部的松罗地区一带。

Xiapu County is located in the south of Fuding City (Fig. 3-36). The phoenix costume in this area is known as the phoenix tail style because of its distinctive phoenix tail bun (Fig. 3-37). The phoenix tail style is not only popular in the She villages of Xiapu County, but also Songluo in the southeast of Fu'an City.

图3-36 霞浦畲乡风光[1]

Fig. 3-36 Scenery of the She village in Xiapu

[1] 图片来源：2018年5月7日摄于福建省宁德市霞浦县溪南镇半月里畲族村。
Source: taken in She Village, Xinan Town, Xiapu County, Ningde City, Fujian Province on May 7, 2018.

图3-37 穿着凤尾式畲族服饰的“中国畲娃”卡通形象

Fig. 3-37 Cartoon of "Chinese She Girls" in phoenix tail style

"凤尾髻"前耸后坠，盘踞成堆，是凤凰形象的异化。它是已婚妇女的发式，梳理起来非常复杂，云髻高鬟中需夹以大量假发（图3-38）。

凤尾式未婚少女的发式则比较简单，通常盘梳成扁圆形，形似红边黑绒帽（图3-39）。

凤尾式凤冠又称"公主顶"，常由笋壳制作，呈金字塔状，缀银流苏（图3-40）。

The front part of the phoenix tail style is high, while the rear part is low. It looks like a bird sitting in a heap, which relates to the phoenix. This style is excluded for married women, which is very complicated to comb and requires a large number of wigs (Fig. 3-38).

Unmarried girls, nevertheless, have a much simpler one. They usually comb the hair into a flat round shape, like a black velvet hat with a red edge (Fig. 3-39).

Phoenix tail-styled crown, also known as princess crown, is made from bamboo shoots. This crown has the shape like a pyramid and is decorated with silver tassels (Fig. 3-40).

图3-38 凤尾髻[1]

Fig. 3-38 Phoenix tail bun

[1] 图片来源：由霞浦县民族与宗教事务局提供。
Source: Provided by Xiapu Bureau of Ethnic and Religious Affairs.

图3–39 凤尾式少女发式
Fig. 3-39 Phoenix tail-styled crown for unmarried women

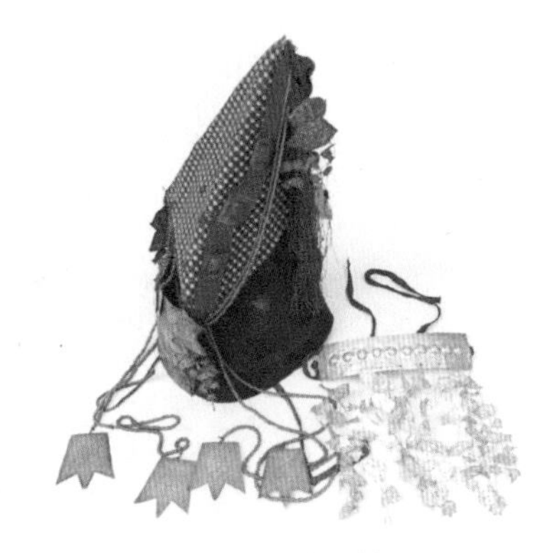

图3–40 凤尾式新娘婚礼凤冠
Fig. 3-40 Phoenix tail-styled crown for bride

凤尾式凤凰装面料颜色以黑（青）色或蓝色为主色调，以少量白色及其他颜色作为配色。它在衣领和前襟处与凤翎式凤凰装颇为相似，其自身特色是前襟绣花为"厂"字形平行排列，且按绣花的组数，分别称为"一红衣""二红衣""三红衣"（图3–41~图3–43）。除此之外，"凤尾式"上衣还可正反两面穿用，居家时将正面朝内以保护绣花，会客时正面朝外华美亮相（图3–44）。

The fabric color of the phoenix tail style is mainly black (cyan) or blue, with a small amount of white and other colors as color matching. Its collar and front are similar to that of the phoenix feather style. Phoenix tail style is mainly featured by the L-shaped front embroideries. And according to the number of embroidery on the front, Single-red garment (one L), Double-red garment (two L in parallel) and Triple-red garment (three L in parallel) are differentiated (Fig. 3-41 to Fig. 3-43). Besides, the phoenix tail type garment can be reversible as you can wear it inside out. They usually wear inside out at home to protect the embroidery, while wearing the opposite to show the gorgeous embroideries in public (Fig. 3-44).

图3–41 凤尾式一红衣[1]

Fig. 3-41 Single-red garment in phoenix tail style

图3–42 凤尾式二红衣[2]

Fig. 3-42 Double-red garment in phoenix tail style

图3–43 凤尾式三红衣[3]

Fig. 3-43 Triple-red garment in phoenix tail style

---

❶ 图片来源：2001年收藏于浙江省博物馆，征集于福安市松罗乡后洋村雷金妹，衣长75厘米，通袖长127厘米。

Source: collected by Lei Jinmei from Houyang Village, Songluo Township, Fu'an City, stored in Zhejiang Provincial Museum in 2001. Total length: 75 cm, sleeve length: 127 cm.

❷ 图片来源：由霞浦县民族与宗教事务局提供。

Source: provided by Xiapu Bureau of Ethnic and Religious Affairs.

❸ 图片来源：2010年4月22日笔者摄于福建省宁德市霞浦县溪南镇白露坑行政村半月里畲族村雷国胜村主任家。

Source: taken at the family of Lei Guosheng, the director of Bailukeng Administrative Village, Xinan Town, Xiapu County, Ningde City, Fujian Province on April 22, 2010.

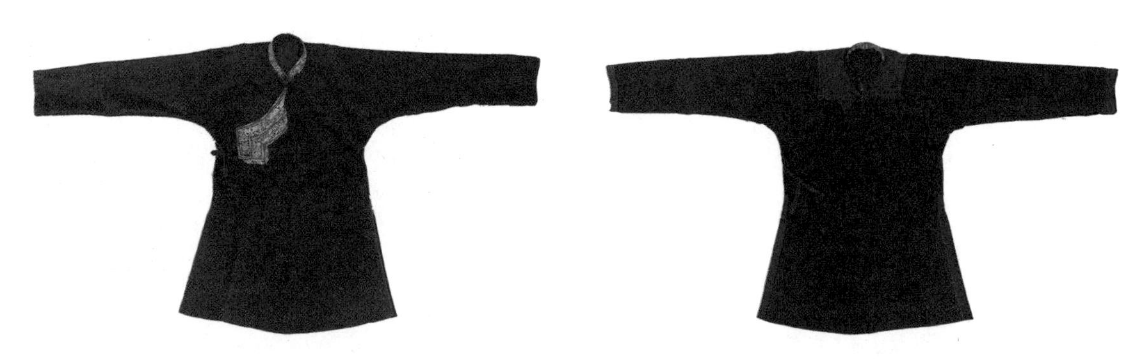

图3-44　凤尾式上衣正反面穿着示意图[1]

Fig. 3-44　The face and reverse side in phoenix tail style

凤尾式围裙裙身呈梯扇形（下端宽且呈弧形）或长方形，常刺绣人物、花、鸟等纹样，非常精美（图3-45）。

The body of the phoenix tail apron features a ladder fan (the lower end is wide and curved) or a rectangle. Embroidered with figures, flowers and birds and other patterns, it looks very exquisite (Fig. 3-45).

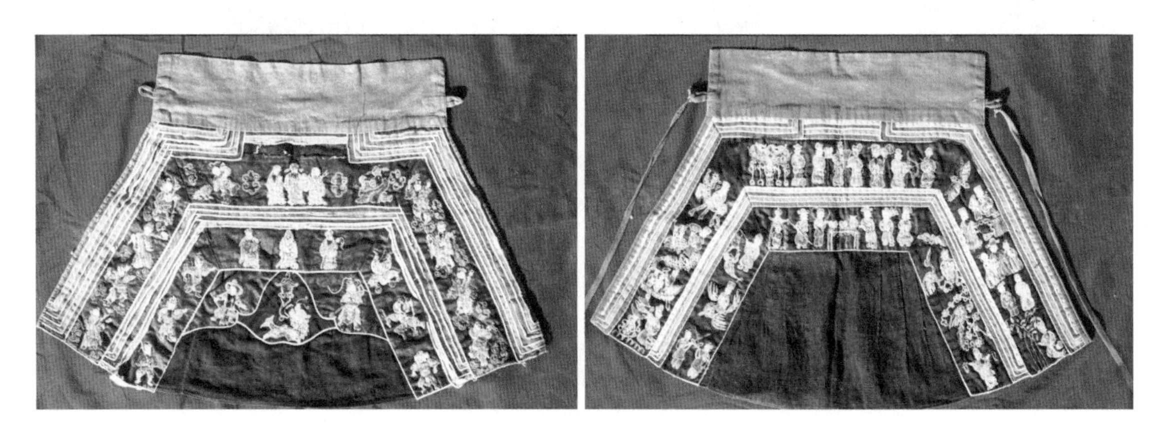

图3-45　凤尾式围裙

Fig. 3-45　Aprons in phoenix tail style

[1] 图片来源：2018年5月7日笔者摄于宁德市霞浦县溪南镇白露坑行政村半月里民间畲族博物馆。
Source: photos were taken in the Museum of the She Nationalities in Banyueli, Bailukeng Village, Xinan Town, Xiapu County, Ningde City, Fujian Province on May 7th, 2018.

凤尾式尖顶花斗笠做工特别精细，俗称“畲家笠斗”，亦称“花笠”。斗笠直径约38厘米，窝深约8厘米，顶高约3厘米，重量只有普通斗笠的一半或三分之二；面层用篾多达224~240条，篾条宽度只1毫米左右，编成的斗笠星（孔）只有5~7毫米，最细的放不下一粒谷子（图3-46）。

The bamboo hat with a conical crown in phoenix tail style is of particularly excellent workmanship, commonly known as the "She hat", also known as the "flower hat". The diameter of the hat is about 38 cm, the depth is about 8 cm, and the height of the crown is about 3 cm. The weight of the hat is only half or two-thirds of that of the ordinary hat. The surface layer uses up to 224~240 thin bamboo strips with the width of each strip within 1 mm roughly. Each hole between strips is only 5 to 7mm, leaving no space for even a grain of millet (Fig. 3-46).

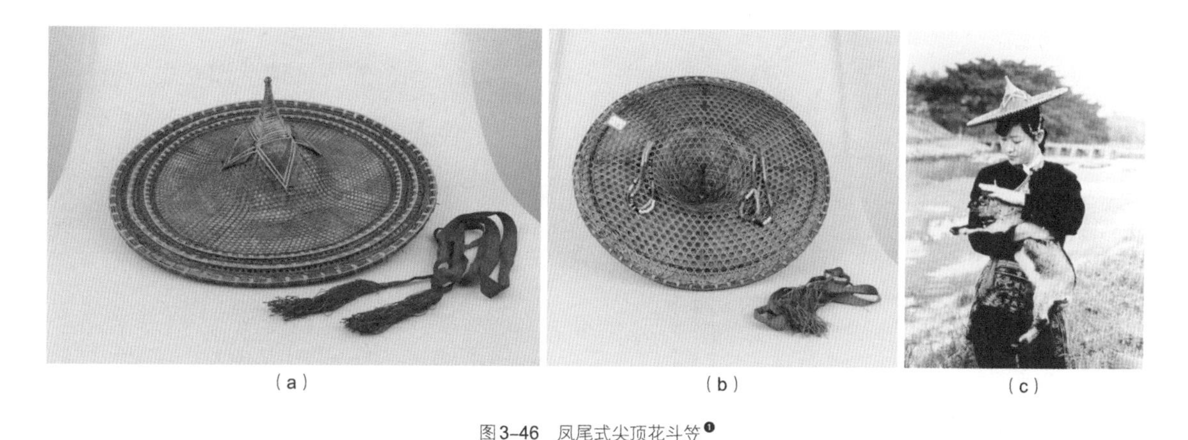

(a) (b) (c)

图3-46 凤尾式尖顶花斗笠[1]

Fig. 3-46 The bamboo hat with a conical crown in phoenix tail style

[1] 图片来源：(a)、(b)：2001年征集于福安市松罗乡后洋村雷金妹，现收藏于浙江省博物馆，民国制作，直径40厘米，高13厘米。(c)：由汤瑛女士提供。

Source: (a),(b): Collected from Lei Jinmei, Houyang Village, Songluo Township, Fu'an City in 2001; stored in Zhejiang Provincial Museum. It was made in Republic of China, with a diameter of 40 cm and a height of 13 cm. (c): Courtesy of Ms. Tang Ying in 2010.

练习题 | Exercise

请为畲族服饰填上你喜欢的色彩吧!

Please color in picture for the She costumes.

五、凤中式

5 Phoenix Trunk Style

福安是福建畲族人口最多、最集中的一个市（图3-47）。该地区畲族妇女多梳“凤身髻”，俗称“凤凰中”，因此我们把这种凤凰装称为“凤中式”。“凤中式”除了流行于福安市外，也流行于周宁县、寿宁县和宁德市的蕉城区八都镇一带（图3-48~图3-50）。

Fu'an has the largest and most concentrated population of the She minority in Fujian Province (Fig. 3-47). In this area, the She women often wear a phoenix trunk bun, commonly known as the middle part of the phoenix (phoenix trunk), so we name it the phoenix trunk style. This kind of phoenix costume is also popular in Zhouning County, Shouning County and Badu Town area in Jiaocheng District, Ningde City(Fig. 3-48 to Fig. 3-50).

图3-47 福安畲乡风光[1]

Fig. 3-47 Scenery of the She village in Fu'an

[1] 图片来源：2010年4月16日笔者摄于福建省福安市社口镇牛山湾村。
Source: Photos were taken by the author in Niushawan Village, Shekou Town, Fu'an City, Fujian Province on April 16, 2010.

图3–48 穿着凤中式畲族服饰的“中国畲娃”卡通形象

Fig. 3-48 Cartoon of "Chinese She Girls" in phoenix trunk style

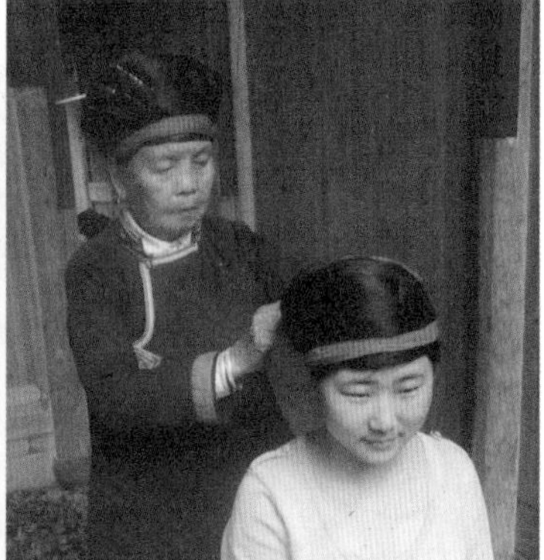

图3-49 福安"凤身髻"[1]

Fig. 3-49 Fu'an phoenix trunk bun

图3-50 凤中式少女（16岁以下）发式[2]

Fig. 3-50 Phoenix trunk-styled crown for She girls (under 16 years old)

---

[1] 图片来源：2010年4月17日笔者摄于福建省宁德市蕉城区八都镇猴盾村。

Source: taken by the author in Houdun Village, Badu Town, Jiaocheng District, Ningde City, Fujian Province on April 17, 2010.

[2] 图片来源：2010年4月16日笔者摄于福建省福安市社口镇牛山湾村。

Source: taken by the author in Niushan Bay Village, Shekou Town, Fu'an City, Fujian Province on April 16, 2010.

“凤中式”凤冠冠身以竹笋壳为骨架，外用红布包缠后缝成长方形头冠。凤冠戴在头上，珠串和银链牌遮住脸部，直垂到胸前，走动起来摇摇晃晃，叮当作响，寓意“凤凰带仔又带孙”。凤冠是畲家姑娘陪嫁礼物之一，在其逝世时，还要戴它入殓。

“凤中式”凤冠大多为如图3–51所示的前高后低的斜顶造型，而在宁德八都地区为如图3–52所示平顶造型。

The body of the phoenix trunk is framed by bamboo shells, with the crown wrapped in red cloth and sewn into a square shape. Placing the phoenix crown on the head, beads and silver chains cover the face and hang straight down to the chest. When walking, it shakes and jingles, implying that the phoenix brings sons/daughters as well as grandsons/granddaughters. As one of the dowry gifts for the She girls, the phoenix crown keeps their company for good, even being buried with them.

Most of the phoenix crown in phoenix trunk style has a slanted top as shown in Fig. 3-51, while in Badu district of Ningde, it has a flat top as shown in Fig. 3-52.

图3–51 福安式凤冠[1]

Fig. 3-51 Phoenix crown in Fu'an style

[1] 图片来源：收藏于景宁畲族博物馆。征集于福建省福安市朝阳新村，民国制造，重350克。

Source: Housed in the She Museum of China. Collected in Chaoyang New Village, Fu'an, Fujian Province. It was made in Republic of China, weighing 350g.

图3-52　福安式宁德八都凤冠[1]

Fig. 3-52　Phoenix crown in Fu'an style in Badu, Ningde

"凤中式"服装具有古朴之美，其面料崇尚蓝黑色、花纹简约稳重。只有在衣领上用水红、黄、大绿等颜色的花线绣虎牙花纹，并沿右衽大襟边镶滚一条3~4厘米宽的红布边。它最突出的特色是上衣右襟下端系带处有一块绣花的角隅花纹。这块三角形花纹据说是模仿当年高辛帝赐给盘瓠王的封印的一半而制（图3-53）。

凤中式盛装围裙一般为黑地红边的梯形，裙身靠裙头处左右各绣一对对称图案，图案凭个人的喜好，龙凤花鸟，各显巧手。一般多为凤凰或牡丹等（图3-54）。

The phoenix trunk style is simple but beautiful, with its color represented by blue and black. The pattern is simple and stable. The collar is embroidered with a tiger's teeth in colored threads of red, yellow, and green. Along the edge of the right front is a red border, spanning 3 to 4 cm. Its most prominent feature is an embroidered corner pattern at the lower right front. The triangular pattern is said to be made in imitation of half of the seal given to King Panhu by Emperor Gaoxin (Fig. 3-53).

The apron in phoenix trunk style is generally a trapezoid black cloth with a red edge. There are a pair of symmetrical patterns embroidered near the apron's waist. The patterns are based on personal preferences, dragons, phoenixes and birds all showing their skillfulness. Under most cases, patterns are featured by phoenix and peony (Fig. 3-54).

[1] 图片来源：福安后舍凤冠，由雷村村主任保存。

Source: Phoenix crown kept by Director of Lei Village in Houshe Village, Fu'an City.

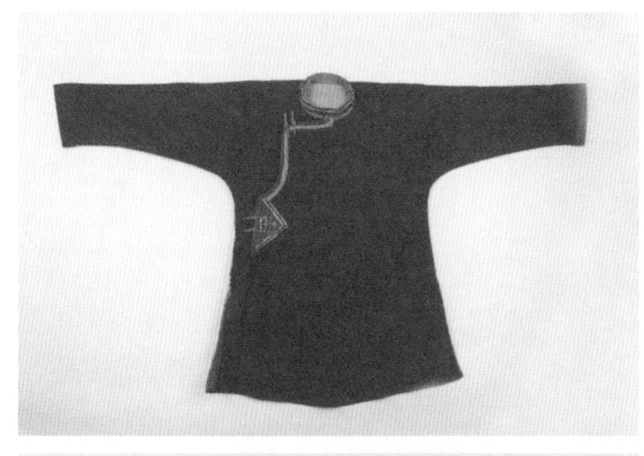
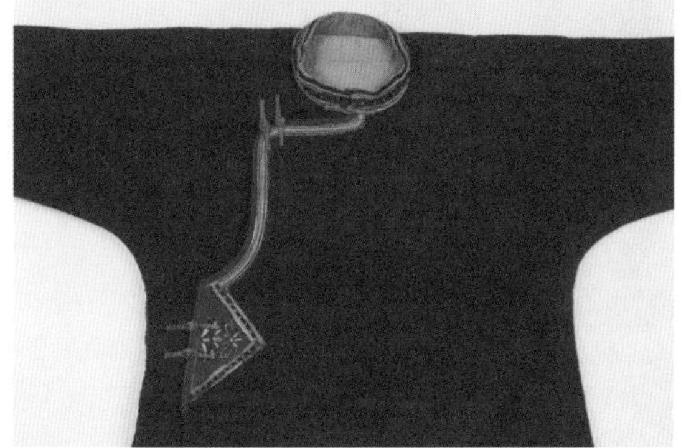

图3–53 凤中式上衣[1]

Fig. 3-53 The garment in phoenix trunk style

图3–54 凤中式凤凰装[2]

Fig. 3-54 Phoenix costume in phoenix trunk style

[1] 图片来源：征集于福安市朝阳新村钟阿青，2001年收藏于浙江省博物馆，民国时期制作，衣长78.5厘米，通袖长132.7厘米，胸围96厘米，底宽65.5厘米，领宽9厘米，领深9厘米。

Source: Collected from Zhong A-qing of Chaoyang New Village in Fu'an City, and stored in Zhejiang Provincial Museum in 2001. It was made in Republic of China. Total length: 78.5 cm, sleeve length: 132.7 cm, chest circumference: 96 cm, bottom width: 65.5 cm, collar width: 9 cm, collar depth: 9 cm.

[2] 图片来源：2010年4月17日笔者摄于福建省宁德市蕉城区八都镇猴盾村。

Source: taken by the author in Houdun Village, Badu Town, Jiaocheng District, Ningde City, Fujian Province on April 17, 2010.

扩展阅读
Further Information

发展中的凤中式凤凰装（图3-55、图3-56）
The Development of the Phoenix Costume in Phoenix Trunk Style (Fig. 3-55, Fig. 3-56)

图3-55　改良过的凤中式畲女服[1]
Fig. 3-55　Modified She garment in phoenix trunk style

[1] 图片来源：2010年4月16日笔者摄于福建省福安市社口镇牛山湾村。
Source: taken by the author in Niushan Bay Village, Shekou Town, Fu'an City, Fujian Province on April 16, 2010.

图3–56 现代凤中式刺绣纹样[1]

Fig. 3-56 Modern embroidery patterns in phoenix trunk style

[1] 图片来源：2010年4月19日摄于福建省福州市罗源县松山镇竹里村国家级畲族“非遗”传承人兰曲钗师傅处。
Source: taken from Master Lan Quchai, the national inheritor of the intangible cultural heritage of She ethnic group, in Zhuli Village, Songshan Town, Luoyuan County, Fuzhou City, Fujian Province on Apr. 19, 2010.

练习题 | Exercise

请为畲族服饰填上你喜欢的色彩吧！

Please color in picture for the She costumes.

六、凤头式

6 Phoenix Head Style

福州罗源、连江地区畲族女装色彩纹饰最为绚烂（图3–57）。宁德市蕉城区南部的飞鸾镇与福州市罗源县相邻，其服饰也与之相似。该式发饰常被称为“凤凰头”，状若优雅颔首的丹顶凤头，因此称为“凤头式”（图3–58~图3–62）。

The color and decoration of the She women's costumes in Luoyuan and Lianjiang counties of Fuzhou, Fujian are the most gorgeous(Fig. 3-57). Feiluan Town, situated in the south of the Jiaocheng District of Ningde City, is adjacent to Luoyuan County of Fuzhou, so the costumes in the two areas share similarities. The hair decoration of this style is often called "phoenix head" as it looks like a red head of an elegant phoenix (Fig. 3-58 to Fig. 3-62).

图3–57　位于罗源的畲服传承示范基地[1]

Fig. 3-57　She costume inheritance demonstration base located in Luoyuan

[1] 图片来源：2018年1月20日摄于福建省福州市罗源县竹里村。
Source: taken in Zhuli Village, Luoyuan County, Fuzhou City, Fujian Province on January 20, 2018.

图3-58 穿着凤头式畲族服饰的“中国畲娃”卡通形象

Fig. 3-58 Cartoon of "Chinese She Girls" in phoenix head style

（a） （b）

图3-59 凤头式少女头饰（16岁以下）[1]

Fig. 3-59 The headwear of She ladies (under 16 years old) in phoenix head style

（a） （b） （c）

图3-60 三种凤头式成年妇女头饰[2]

Fig. 3-60 Three types of adult women's headwear in phoenix head style

---

[1] 图片来源：（a）：笔者2010年4月18日摄于宁德市蕉城区飞鸾镇南山村；（b）：2010年由汤瑛女士提供。

Source: (a): taken in Nanshan Village, Feiluan Town, Jiaocheng District, Ningde City on April 18, 2010; (b): Courtesy of Ms. Tang Ying in 2010.

[2] 图片来源：（a）：由福建省福州市连江县东湖镇天竹村提供；（b）：笔者2018年1月21日摄于福建省福州市连江县小沧畲族乡；（c）：笔者2010年4月18日摄于宁德市蕉城区飞鸾镇南山村。

Source: (a): Provided by Tianzhu Village, Donghu Town, Lianjiang County, Fuzhou City, Fujian Province; (b): Taken in Xiaocang She ethnic group Township, Lianjiang County, Fuzhou, city Fujian Province on January 21, 2018; (c): Taken in Nanshan Village, Feiluan Town, Jiaocheng District, Ningde City on April 18, 2010.

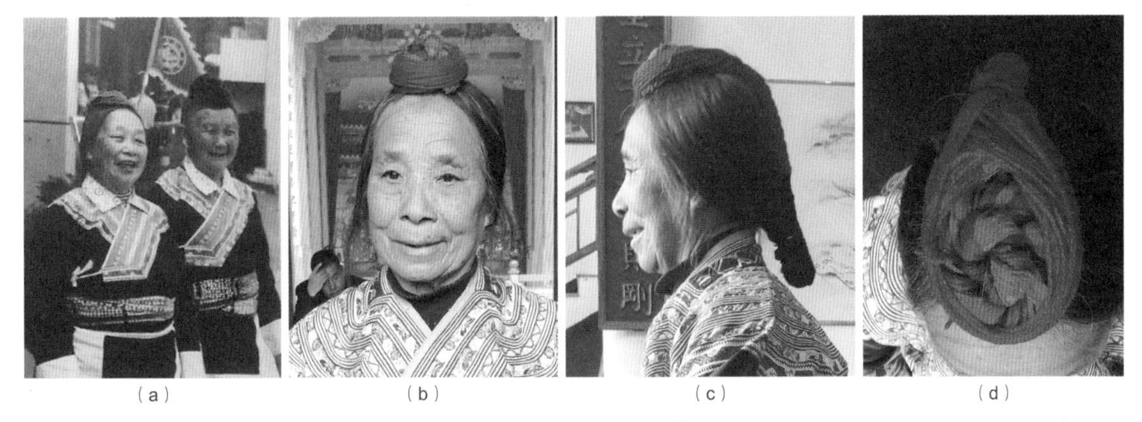

（a） （b） （c） （d）

图3-61 凤头式老年发式[1]

Fig. 3-61 Senior women's headwear in phoenix head style

图3-62 凤头式新娘凤冠[2]

Fig. 3-62 Bride headwear in phoenix head style

[1] 图片来源：（a）：由福建省福州市连江县东湖镇天竹村村委会提供；（b）～（d）：为笔者于2018年1月21日摄于该村。

Source: (a): provided by the villagers' committee of Tianzhu village, Donghu town, Lianjiang county, Fuzhou city, Fujian province; (b) to (d): Photographed in Tianzhu Village on January 21, 2018.

[2] 图片来源：汤瑛女士提供。

Source: courtesy of Ms. Tang Ying in 2010.

罗源在畲族自南向北的迁徙过程中是一个关键中转站，可以说是除祖居地广东凤凰山之外的畲族第二发源地。该地区的凤凰装保留了大量的文化信息。当地畲家人这样生动形象地介绍他们的“凤凰装”：幼年、成年、老年的不同妇女头式，是依小凤凰、大凤凰和老凤凰的模样打扮的。而且服装各部分都有象征意义。如已婚青年妇女的装扮，头饰象征凤冠，衣领、襟、袖所绣花边分别象征凤凰的颈、腰和翅膀，围裙象征凤腹，身后的两条飘带象征凤尾，花绑腿及绣花鞋象征凤爪（图3-63、图3-64）。

Luoyuan plays a vital role in the migration of the She ethnic group from the south to the north. It is hailed as the second birthplace of the She ethnic group, while Fenghuang Mountain in Guangdong being the first one. The phoenix costume of this region retains plenty of cultural genes. The local She people vividly introduce their phoenix costume: women of different ages have different head styles. The tenders, adults, and seniors dress themselves up by mimicking little, grown-up and senior phoenixes respectively. And every part of the costume has symbolic meaning. For example, young married women's headdress symbolizes the phoenix crown, the embroidered edges of the collar, lapel and sleeve respectively symbolize the neck, waist and wings of the phoenix. The apron symbolizes the phoenix belly; the two streamers behind it symbolize the phoenix tails; and the flower gaiters and embroidered shoes symbolize phoenix feet(Fig. 3-63, Fig. 3-64).

（a）罗源凤头式

(a) Luoyuan style

（b）连江凤头式

(b) Lianjiang style

（c）飞鸾凤头式

(c) Feiluan style

图3-63　凤头式畲族盛装[1]

Fig. 3-63　She costume in phoenix head style

[1] 图片来源：（a）：为汤瑛女士提供；（b）、（c）：由福建省福州市罗源县霍口乡福湖村文化礼堂提供。

Source: (a): Courtesy of Ms. Tang Ying in 2010; (b),(c): Provided by Fuhu Village Cultural Auditorium, Huokou Town, Luoyuan County, Fuzhou City, Fujian Province.

图3-64 凤头式畲族盛装[1]

Fig. 3-64 She costume in phoenix head style

[1] 图片来源：笔者于2018年1月21日摄于福建省福州市连江县东湖镇天竹村。

Source: taken in Tianzhu Village, Donghu Town, Lianjiang County, Fuzhou City, Fujian Province on January 21, 2018.

扩展阅读
Further Information

历史上的罗源畲族服饰
Luoyuan She Costumes in History

乾隆时期《皇清职贡图》中所绘罗源畲族男女服装款式为大襟右衽（图3-65），同于当地汉民。

人类学家凌纯声先生描述民国时福州罗冈式畲族妇女头饰由前中后明显的三部分组成，这与伊莎贝拉（Isabella L.）根据1898年游历中国见闻所撰写的《长江流域及其腹地》一书中所引照片（图3-66）一致。前部为一彩穗流苏；中部是在一个覆盖着银片的筒形上方再架起一座三角形布帐；后部为披着一片矩形布片的发簪，插入中部下的发髻内。书中提及该照片摄于福州以北40英里处，即约62公里，推断为福建省罗源县霍口畲族乡附近。

As per records in *Illustrations of Tributary Peoples* which was completed during Qianlong's reign (1736-1796), costumes of the She men and women in Luoyuan features big garment with right lapel (Fig. 3-65), the same as those of the local Han people.

Mr. Ling Chunsheng described that in Republic of China, the She women's headwear in Luogang Fuzhou style consisted of three distinct parts, the front, the middle and the rear part. This description echoed with photos cited by Isabella L. in *The Yangtze Valley and Beyond* (1898) (Fig. 3-66): a colorful tassel on the front; a triangular cloth tent set up above a cylinder covered with silver sheets in the middle; a hairpin draped with a rectangular piece of cloth, inserted into the bun at the back. Isabella's book mentioned that the photo was taken 40 miles (about 62km) north of Fuzhou. It can be referred to that the place is near Huokou She Township, Luoyuan County, Fujian Province.

图3–65 清代《皇清职贡图》载福建罗源畲族服饰

Fig. 3-65 Luoyuan She costumes recorded in *Illustrations of Tributary Peoples* in Qing Dynasty

20世纪初，美国著名旅行家威廉·埃德加·盖洛（William Edgar Geil）于清末对中国的十八个省府进行了广泛而细致的考察，并著《中国十八个省府》。如图3-67所示，书中福州部分老照片呈现了当时福州畲族妇女的形象。中间两位畲族妇女下身穿着紧身及膝中裤，精干爽利。

At the beginning of the 20th century, William Edgar Geil, a famous American traveler, made an extensive and detailed field study of the 18 provinces on China's mainland at the end of the Qing Dynasty and completed the *Eighteen Capitals of China*. As shown in Fig. 3-67, some old photos of Fuzhou in the book present the images of the She women in Fuzhou at that time. The two She women in the middle are wearing tight knee-length trousers, looking dapper and smart.

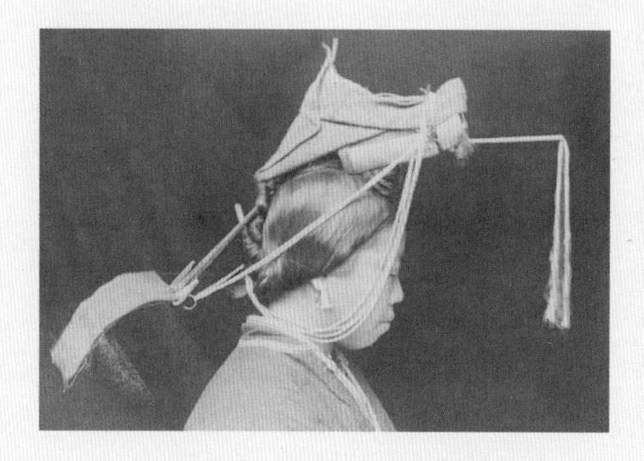

图3-66 19世纪末福州畲族头饰[1]

Fig. 3-66 She woman hair accessory in Fuzhou by the end of 19th century

图3-67 清末民初罗源式服饰[2]

Fig. 3-67 She costume in Luoyuan style in late Qing Dynasty and early Republic of China

[1] 图片来源/Source：Isabella L. The Yangtze Valley and Beyond[M]. Cambridge: Cambridge University Press, 2010.

[2] 图片来源/Source: William Edgar Geil. Eighteen Capitals of China[M]. Philadelphia: Washington Square Press, 1911.

练习题 | Exercise

请为畲族服饰填上你喜欢的色彩吧!

Please color in picture for the She costumes.

七、扇屏式

7 Fan-shaped Style

扇屏式凤凰装主要流行于福建省南平市顺昌县境内（图3-68），穿着该式服装的人口在8000人左右，约占畲族总人口的1%。如图3-69所示，其扇形头饰造型独特，形似孔雀伞状冠羽，具有浓厚的地方特色。

The phoenix costume in fan-shaped style is mainly popular in Shunchang County, Nanping City, Fujian Province(Fig. 3-68). About 8000 people there are wearing this style, accounting for one percent of the total population of the She ethnic group. As shown in Fig. 3-69, its fan-shaped headdress has a unique shape, resembling a peacock's umbrella-shaped feather, which has strong local characteristics.

图3–68　顺昌畲族民居❶

Fig. 3-68　She house in Shunchang

❶ 图片来源：2018年1月16日笔者摄于福建省南平市顺昌县岚下乡桃源村。

Source: taken by the author in Taoyuan Village, Lanxia Township, Shunchang County, Nanping City, Fujian Province on Jan 16, 2018.

图3–69　穿着扇屏式畲族服饰的“中国畲娃”卡通形象

Fig. 3-69　Cartoon of "Chinese She Girls" in fan-shaped style

扇屏式头饰在当地也被称为“盘瓠顶”（图3-70）。其中，顺昌式头饰的主要配件为由几十甚至上百支（图3-71为58支）银簪排成的扇形银饰“头笄”（图3-72）。

The headdress in fan-shaped style is termed as Panhu style by local people(Fig. 3-70). As shown in Fig. 3-71, the main accessories of Shunchang style headdress is silver touji, consisting of dozens or even hundreds of silver hairpins (58 pieces in Fig. 3-72) arranged in a fan shape.

图3-70 扇屏式畲族妇女头饰[1]

Fig. 3-70 The headwear of married She women in fan-shaped style

[1] 图片来源：笔者2018年1月15日摄于福建省南平市顺昌县畲族文化研究会。

Source: taken by the She Culture Research Association in Shunchang County, Nanping City, Fujian Province on January 15, 2018.

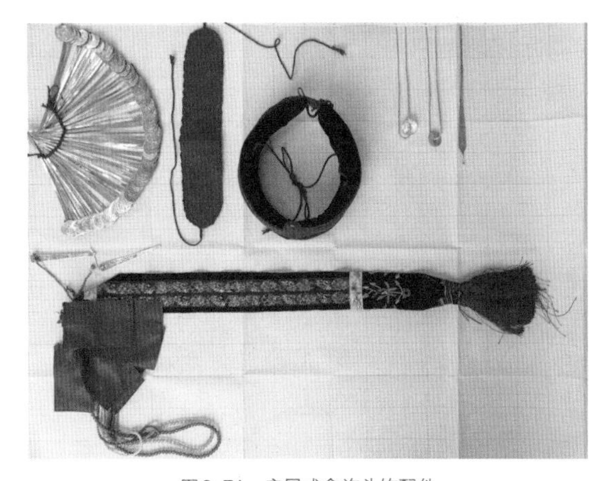

图3-71 扇屏式畲族头饰配件

Fig. 3-71 Headwear in fan-shaped style

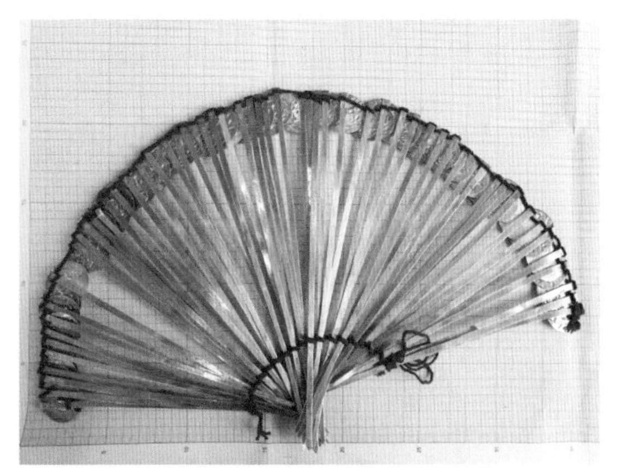

图3-72 扇屏式畲族扇形银头饰“头笄”反面

Fig. 3-72 Back of silver touji in fan-shaped style

扇形银饰中的银簪如图3-73所示，每支长约18厘米，前端折回呈钩形，上面的圆片宽2.8厘米，长2.5厘米，正面刻有凤穿牡丹、观音送子、仙鹤延年、福禄（鹿）高升（竹）和熊猫等纹样（图3-74）。过去发簪可代表家庭贫富，富者戴大量的银簪，贫者戴少量的铜簪。

The silver hairpins in the fan-shaped silver ornaments are shown in Fig. 3-73. Each hairpin is about 18cm long, with a hook-shaped front end connecting to a round silver wafer. The wafer is 2.8cm wide and 2.5cm long. The front side of the wafer is engraved with the patterns of the phoenix and the peony, a Buddism goddess Guanyin, crane, deer, bamboo and panda (Fig. 3-74), hoping to bring fortunes, fame, longevity and offspring. In the past, numbers of hairpins told the family backgrounds: the rich wore many silver hairpins while the poor wore few copper hairpins.

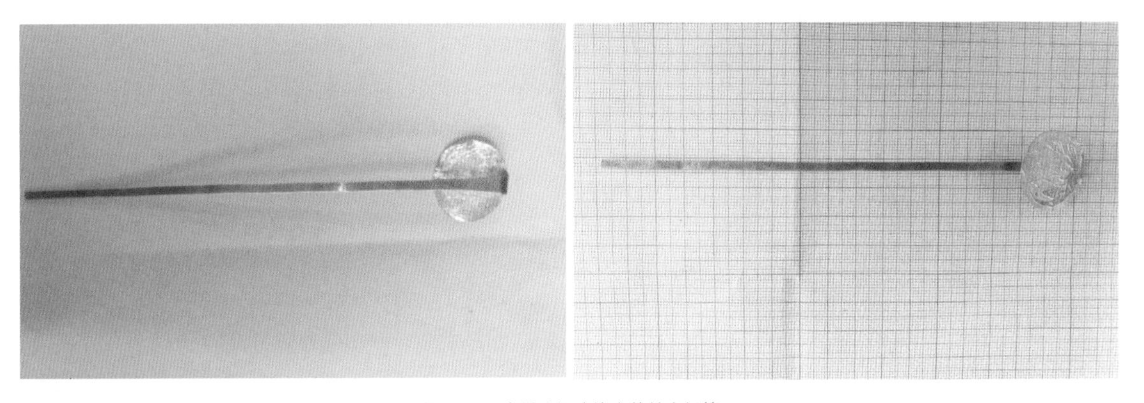

图3-73 扇屏式银头饰中的单支银簪

Fig. 3-73 Single hairpin in the fan-shape silver headwear

图3-74 扇屏式银头饰中银簪上的图案

Fig. 3-74 The patterns on the hairpins in the fan-shaped silver headwear

扇屏式服装主要包括大襟衣、裙子、绑腿、花鞋等。上衣为右衽大襟，微领，衣身较宽大，前后裾等长，袖口和衣衩内缘绲红色布边，衣领和大襟角绣有简单的纹饰（图3-75）。裙子为过膝黑裙，上沿有白边，裙身两侧边缘绲有红绿布边。绑腿为白色梯形，上有红色和黄色的系带。花鞋为自纳黑色布鞋，鞋口边缘缀红边，鞋头前凸，鞋脊起棱。

The She costume in fan-shaped type mainly features a large garment, skirt, puttee, flowery shoes, etc. The garment has right lapel and a short collar. The front and back sides of the wide garment are of equal length. The cuffs and vents are sewed with red edges. There are simple patterns on the collar and the edges of the garment(Fig. 3-75). The skirt is black and above the knee, with a white hem on the top and a red and green hem on both sides. The puttees are white in a trapezoidal shape with red and yellow lace. Flowery shoes are made of black cotton and they are homemade, with the shoe mouth decorated with a red edge, the front part protruding, and the ridge standing.

图3-75　扇屏式畲族女服图像[1]

Fig. 3-75　She costume in Fan-shaped style

---

[1] 图片来源：笔者2018年1月16日翻拍于溪南村大坪22号。图片中女子为雷路妹。画像于1971年绘制。
Source: photo was re-shot in Xinan Village on January 16, 2018. The woman in the picture is Lei Lu mei. Picture was drawn in 1971.

自20世纪50年代后，当地畲民就很少穿着传统民族服饰。以前结婚装并非红色，而是上衣为蓝色，下面配黑色的大摆裙。只有新娘坐轿子时，才会在头上铺一个六尺的红布。婚礼时也会跳马灯舞或火把舞。现代扇屏式畲族新娘服饰如图3-76所示，胸前配12厘米宽八卦铜牌。平日盛装多为图3-77款式。

After 1950s, the local She people rarely wore their traditional ethnic costumes. In the past, the wedding dress was not red, instead, the upper part was blue, with a big black skirt under it. Only when the bride is in a sedan chair on the wedding day, a 2m red cloth will be spread over her head. There was also a horse dance or a torch dance at weddings. Modern She costume for bride in fan-shaped style is shown in Fig. 3-76, from which we can also find an eight diagrams bronze plate with a width of 12 cm. Formal costume in this style is shown in Fig. 3-77.

图3-76 现代扇屏式畲族新娘服[1]

Fig. 3-76 Modern She costume for bride in fan-shaped style

图3-77 现代扇屏式畲族女服

Fig. 3-77 Modern She costume in fan-shaped style

[1] 图片来源：由雷弯山教授提供。

Source: Courtesy of Professor Lei Wanshan.

练习题 | Exercise

请为畲族服饰填上你喜欢的色彩吧！

Please color in picture for the She costumes.

八、耳巾式

8 Ear-shaped Scarf Style

江西贵溪市樟坪畲族乡（图3–78）的凤凰装以其形似狗耳的头巾为特色，当地畲家人称为“狗耳巾”，因此其凤凰装服式被称为“耳巾式”（图3–79、图3–80）。

In the Zhangping She ethnic group township, Guixi city, Jiangxi province(Fig. 3-78), the phoenix costume is characterized by a scarf shaped like dog ears. The local She people call it dog-ear scarf, and that is why we name it ear-shaped scarf style(Fig. 3-79, Fig. 3-80).

图3–78　樟坪畲族民居[1]

Fig. 3-78　She house in Zhangping, Guixi, Jiangxi

[1] 图片来源：笔者2012年8月15日摄于江西省贵溪市樟坪畲族乡。
Source: taken by the author in Zhangping She township, Guixi city, Jiangxi province on Aug.15, 2012.

图3-79 穿着耳巾式畲族服饰的“中国畲娃”卡通形象

Fig. 3-79 Cartoon of "Chinese She Girls" in ear-shaped scarf style

图3–80 “狗耳巾”的穿戴步骤[1]

Fig. 3-80 Steps of wearing dog ear scarf

[1] 图片来源：笔者2012年8月15日摄于江西省贵溪市樟坪畲族乡。

Source: taken by the author in Zhangping She Township, Guixi City, Jiangxi Province on Aug.15, 2012.

耳巾式凤凰装日常服饰与景宁式等大部分畲族地区常服款式相仿。传统畲族服饰特征在女装上有所保留，中老年妇女常着青（黑）色右衽大襟装，袖口、领围、裤脚等处镶有花边。当地畲族妇女还习惯穿着长及小腿的围裙（图3–81）。

The phoenix costume in ear-shaped scarf style for daily wear is similar to that of most She ethnic group areas represented by Jingning. The characteristics of the traditional She clothing are still retained in women's clothing nowadays. Middle-aged and elderly women often wear a blue or black wide garment with right lapel, and lace is inlaid at the cuffs, collar and trouser legs. Local She women are also used to wearing an apron as long as the calf (Fig. 3-81).

(a) (b)

图3–81 耳巾式畲族日常服饰[1]

Fig. 3-81 Daily She costume in ear-shaped style

[1] 图片来源：(a)：笔者2012年8月15日摄于江西省贵溪市樟坪畲族乡；(b)：《贵溪樟坪畲族志》(评审稿)彩图页，朱友林摄。
Source: (a): taken by the author in Zhangping She Township, Guixi City, Jiangxi Province on Aug.15, 2012; (b): *The record of the She ethnic group in Zhangping, Guixi* (Working draft) color page, photo by Zhu Youlin.

扩展阅读
Further Information

历史上的樟坪畲族服饰
Zhangping She Costumes in History

中华人民共和国成立前江西畲族男性扎头巾，已婚妇女梳高头髻，未婚女子梳辫子。中华人民共和国成立后男女同扎狗耳巾（图3-82）。但如图3-83所示，直到1990年，畲族妇女的头饰还是保留有花边巾覆头的形式。

Before the founding of the People's Republic of China in 1949, She men in Jiangxi wore scarves; married women there combed their hair into buns and unmarried women braided. After 1949, both men and women tied dog ear scarves(Fig. 3-82). The difference lied in the lace scarf for women (Fig. 3-83), which lasted till 1990.

图3-82 《贵溪樟坪畲族志》中所载照片资料——男子头饰（1966年前后）
Fig. 3-82 Men headwear in *The Record of She ethnic group in Zhangping, Guixi* (pictured in 1966 or so)

图3-83 《贵溪樟坪畲族志》中所载照片资料——妇女头饰
Fig. 3-83 Women headwear in *The Record of She ethnic group in Zhangping, Guixi*

❶ 图片来源：《贵溪樟坪畲族志》彩图页，王陵波摄。
Source: *The Record of She ethnic group in Zhangping, Guixi* color page, photo shot by Wang Lingbo.

练习题 | Exercise

请为畲族服饰填上你喜欢的色彩吧!

Please color in picture for the She costumes.

九、头包式

9 Headband Style

贵州畲族服饰，因地区不同而略有差异，以占贵州畲族人口绝大多数的麻江县六堡式畲族盛装服饰最具代表性。当地已婚畲族妇女把头发梳向后侧挽成髻团，并用马尾等作成的发网网住发团，插上发簪，包上藏青色长2米（约6尺）、宽33厘米（约1尺）白底蜡染兰花头帕，称“头包”。根据它的头饰特色，我们将麻江六堡凤凰装称为“头包式”（图3-84、图3-85）。

The costumes of the She minority in Guizhou vary slightly from place to place, with the style in Liubao She Village, Majiang County being most representative, as this area boasts the largest She population in Guizhou. Local married She women comb their hair back and pull it up into a bun. They use hair nets made of horsetails to fasten the hair, put hairpins on the hair, and wrap the hair in a navy-blue batik headscarf. Such a headscarf has a length of 2m and a width of 33cm, which is called a headband. As per the characteristics, we call the She costume in this area headband style (Fig. 3-84, Fig. 3-85).

图3-84 头包式畲族妇女头饰[1]

Fig. 3-84 The headwear of She women in headband style

[1] 图片来源：钱仕豪，麻江，畲族同胞欢庆“四月八”，当代先锋网。
Source: Qian Shihao. Majiang, She Compatriots Celebrate "April 8th", Contemporary Pioneer Network.

图3-85　穿着头包式畲族服饰的“中国畲娃”卡通形象
Fig. 3-85　Cartoon of "Chinese She Girls" in headband style

除“头包”外，麻江畲族盛装上衣相比其他畲族服式最有特色之处在于叠穿的上衣和拼接的袖子。

上衣一般为三件、四件或六件叠穿。制作服装时，面料叠放、衣片一次裁剪成形。每件衣服的下摆和衣角均有红、白两条刺绣花边装饰，内长外短，每层外面衣服均比里面一层短约7厘米（2寸），以露出里面衣服衣摆上的花边为宜，以展示刺绣技艺。

袖子由衣身袖和接袖两部分构成，衣身袖与衣身连裁，接袖袖长约52.8厘米（16寸）。接袖由蜡染和刺绣两段组成，蜡染段为白底蓝花，刺绣花纹段分上中下三部分，上下两边多为“寿”字纹、梅花纹，中部为花纹图案，以牡丹、月季为主（图3-86）。

In addition to the headband, the most distinctive features of the She costume in Majiang lie in the multilayer garments and spliced sleeves.

Three, four or six layers of garments are usually stacked and worn together. Fabric is overlapped when making clothing, and the garment will be cut and made in one go. The hems and corners of each garment are decorated with red and white embroidery lace, long inside and short outside. The outer layer of each garment is about 7cm shorter than the inner layer, to expose the lace on the inner hem of the garment in a bit to show the embroidery skill.

The sleeves are made up of two parts, the garment sleeves and the spliced sleeves. The garment sleeves are cut together with the body of the garment. The length of the spliced sleeves is approximately 52.8 cm. The spliced sleeves are made up of batik section and embroidery section. The batik section has blue flowers on a white background. The pattern of the embroidery section is divided into three parts: upper, middle and lower. The upper and lower sides are mostly patterned by the character of "Shou" (longevity) or the shape of plum flowers, while the middle side is patterned by flowers represented by peony and Chinese rose(Fig. 3-86).

图3–86　头包式畲族盛装服饰[1]

Fig. 3-86　She costume in headband style

[1] 图片来源：六堡村中心网站。

Source: center website of Liubao Village.

练习题 | Exercise

请为畲族服饰填上你喜欢的色彩吧!

Please color in picture for the She costumes.

第四章

Chapter

4

精巧秀丽的畲族手工艺

Exquisite She Handicrafts

小畲凤：既好看又实用！

Little Phoenix: Good-looking and practical.

小畲凤：看了这么多种凤凰装，你最喜欢哪一种哇？小畲凤我可是每一种都喜欢得不得了！因为每一种凤凰装美丽的外表下，都藏着我们畲家人独特的设计智慧和高超的工艺技巧呢！下面我们来一一介绍下畲族手工艺。

Little Phoenix: Of the nine phoenix costumes, which one do you like best? I like all of them as each tells the wisdom of She people and their craftsmanship. Now She handicrafts will be shown.

一、织造工艺

1 Weaving

咱们先来聊聊"织"。提到"织"，那可就必须要来说说畲家的围裙。如前文展示，各地围裙在造型、材质、色彩、图案上各具特色。而它们都有一个共同的亮点，那就是两侧作为系带的织带（图4–1、图4–2）。

Let's talk a little about the weaving first. Speaking of weaving, the apron is a must-see. As shown before, apron in different places has its features in shape, material, color, and pattern. All of them share something in common: webbing ties (Fig. 4-1, Fig. 4-2).

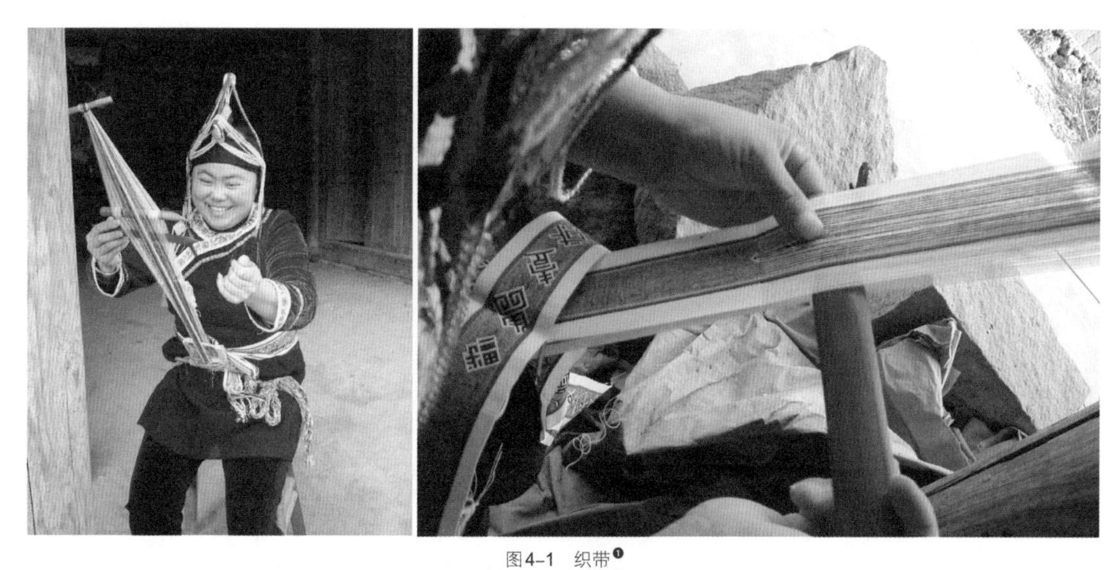

图4-1 织带[1]

Fig. 4-1 Weaving the ribbon

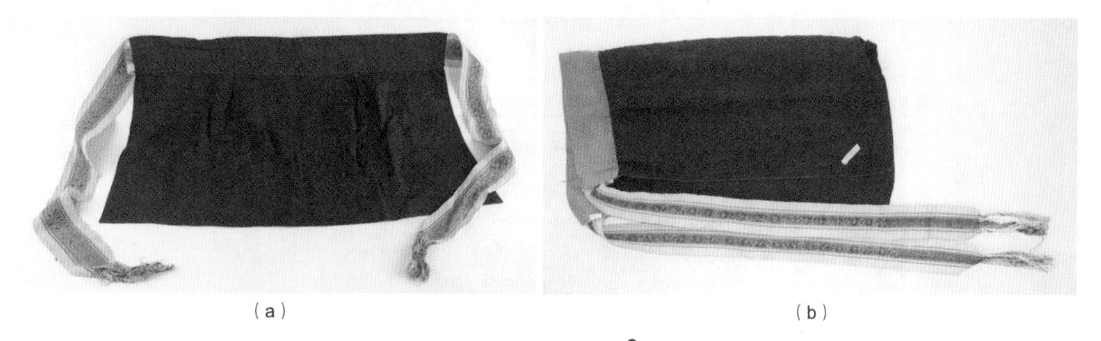

（a） （b）

图4-2 景宁畲族围裙[2]

Fig. 4-2 Apron in Jingning

❶ 图片来源：2004年2月28日笔者摄于浙江省景宁县织带传承人蓝延兰工作室。

Source: taken in the workshop of Lan Yanlan, the inheritor of ribbon-weaving in Jingning County, Zhejiang Province on Feb. 28th, 2004.

❷ 图片来源：（a）：776#二排字宽带彩带青布围裙。征集于景宁大均伏坑雷秀花，2001年收藏于浙江省博物馆，20世纪50年代制作，长200厘米，宽44厘米。（b）：702#真丝围裙。征集于景宁大均伏坑雷秀花，2000年收藏于浙江省博物馆，长182厘米，宽43厘米。

Source: (a): 776# blue cotton apron. It was collected from Lei Xiuhua living in Dajunfu Keng in Jingning and stored in Zhejiang Provincial Museum in 2001. Made in the 1950s, the apron is 200 cm long and 44 cm wide. (b): 702# silk apron. It was also collected from Lei Xiuhua, and collected in Zhejiang Provincial Museum in 2000. It is 182 cm long and 43 cm wide.

可别小看这条小小的手工织带哦，它可是堪称凤凰装中的“活化石”。织带上代代相传的纹样反映着千百年来畲家人从衣食住行到思想观念的方方面面，对于没有本族文字的畲家人来说，就像一个个暗藏着远古信息的密码，有学者认为这些纹样起着象形文字的作用（图4-3~图4-6）！

The webbing ties of the apron, hailed as the living fossil of phoenix costumes, went beyond the item itself. The patterns on the tie, which were handed down from generation to generation, reflect all aspects of the life of the She people from clothing, food, shelter, transportation to ways of thinking over the past thousands of years. To the She people who do not have their own written language, such patterns are like a secret code containing ancient information. Some scholars even believe that these patterns play the role of hieroglyphs(Fig. 4-3 to Fig. 4-6).

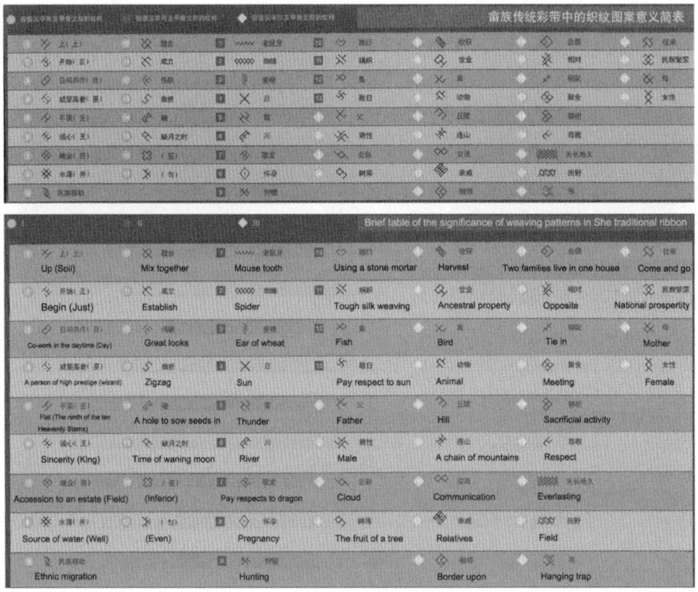

畲族传统彩带中的织纹图案意义简表

Brief table of the significance of weaving patterns in She traditional ribbon

I	II	III				
Up (Soil)	Mix together	Mouse tooth	Using a stone mortar	Harvest	Two families live in one house	Come and go
Begin (Just)	Establish	Spider	Tough silk weaving	Ancestral property	Opposite	National prosperity
Co-work in the daytime (Day)	Great looks	Ear of wheat	Fish	Bird	Tie in	Mother
A person of high prestige (wizard)	Zigzag	Sun	Pay respect to sun	Animal	Meeting	Female
Flat (The ninth of the ten Heavenly Stems)	A hole to sow seeds in	Thunder	Father	Hill	Sacrificial activity	
Sincerity (King)	Time of waning moon	River	Male	A chain of mountains	Respect	
Accession to an estate (Field)	(Inferior)	Pay respects to dragon	Cloud	Communication	Everlasting	
Source of water (Well)	(Even)	Pregnancy	The fruit of a tree	Relatives	Field	
Ethnic migration		Hunting		Border upon	Hanging trap	

图4-3　畲族传统彩带中的织纹图案意义[1]

Fig. 4-3　Meaning of patterns in She traditional webbing ties[1]

[1] 图片来源：景宁中国畲族博物馆。
Source: She Museum of China in Jingning.

图4-4 近代丽水畲族围裙带[1]

Fig. 4-4 Modern webbing tie of apron in Lishui

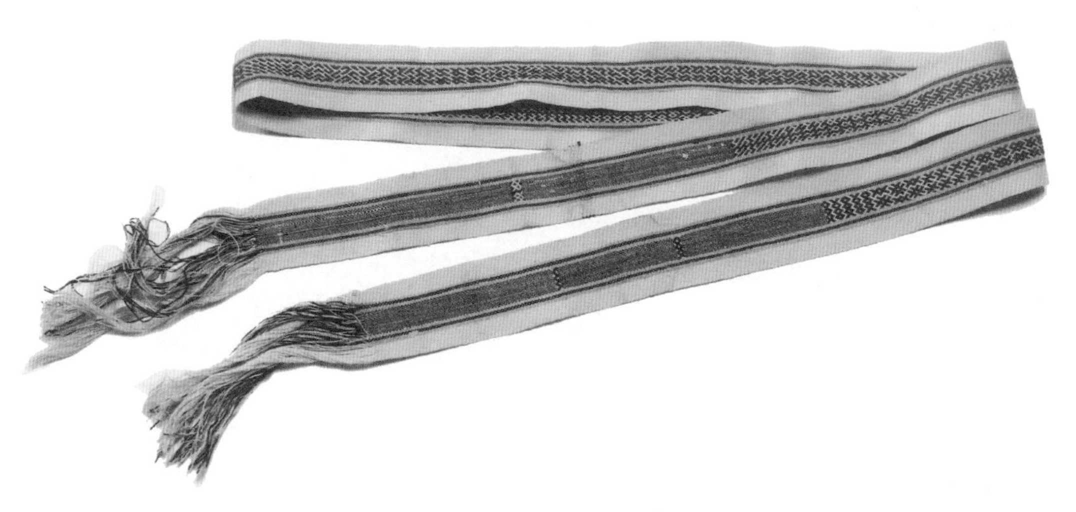

图4-5 近代平阳畲族手织带[2]

Fig. 4-5 Modern handmade webbeing tie in Pingyang

---

[1] 图片来源：征集于丽水市云和县平垟岗村，1959年收藏于浙江省博物馆，长163厘米，宽4.6厘米。

Source: collected from Pingyang Gang Village, Yunhe County, Lishui City. It was stored in Zhejiang Provincial Museum in 1959. It is 163 cm long and 4.6 cm wide.

[2] 图片来源：1959年征集于平阳县山门，现藏于浙江省博物馆，长90厘米，宽2厘米。

Source: collected in Pingyang County Shanmen in 1959, and stored in Zhejiang Provincial Museum. It is 90 cm long, 2 cm wide.

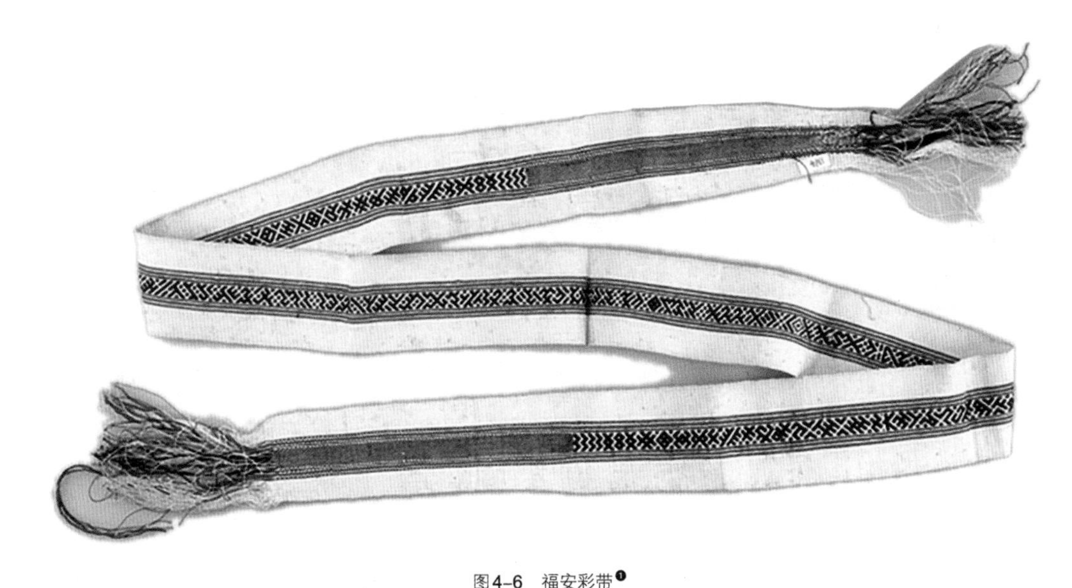

图4-6 福安彩带[1]

Fig. 4-6 Webbing tie in Fu'an

除了传统纹样，心灵手巧的畲家女儿也常手随眼动，灵活撷取身边的文字，反映社会生活的方方面面。图4-7所示的民国时期织带上织有“风调雨顺国泰民安皇帝万寿宋元明清顺治康熙雍正嘉庆乾隆道光咸丰同治……民国龙飞凤舞荣华富贵……”等字样，就像一条缠在腰上的历史书！

Besides traditional patterns, intelligent She girls weave what they see in daily life, thus reflecting all aspects of social life. As shown in Fig. 4-7, the webbing tie in Republic of China period reads "The weather is favorable and the people are in peace. Long live emperors in Song, Yuan, Ming and Qing (Dynasties), i.e. Shunzhi, Kangxi, Yongzheng, Jiaqing, Qianlong, Daoguang, Xianfeng, Tongzhi... Republic of China is prosperous..." and so on. It is just like a history book pasted along the waist!

---

[1] 图片来源：浙江省博物馆。
Source: Zhejiang Provincial Museum.

现代织带上的文字更丰富，有织于中华人民共和国成立之初的“毛主席语录”带（图4–8），还有织有“浙江省景宁畲族自治县迎一九九七香港回归庆贺香港回归1周年纪念”和“浙江省景宁畲族自治县迎一九九九澳门回归庆贺澳门回归祖国”字样的织带（图4–9），表达了畲家人对祖国的关注和支持！

Modern webbing ties reveal rich content. Many of them, such as the one woven with words "quotations from Chairman Mao" made at the beginning of the founding (Fig. 4-8), and another two woven the words as"Jingning She Antonomous County of Zhejiang celebrates the first anniversary of Hong Kong's return in 1997" and "Jingning She Antonomous County of Zhejiang celebrates the Macao's return to the motherland (Fig. 4-9). All of these expressed the She people's great attention and support for the motherland.

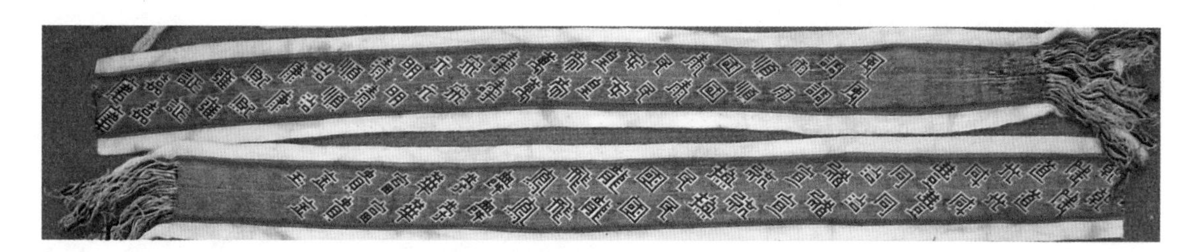

图4–7 景宁畲族织带

Fig. 4-7 Webbing tie in Jingning

图4–8 景宁畲族织带“毛主席语录”

Fig. 4-8 Webbing tie in Jingning"reading Quotations from Chairman Mao"

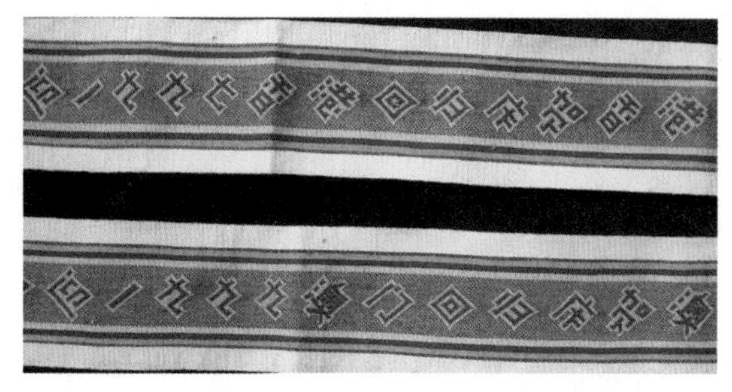

图4–9 景宁畲族织带——迎港澳回归

Fig. 4-9 Webbing ties in Jingning, welcoming the return of Hong Kong and Macao

这些精美的织带还常被畲家姑娘用来作为爱情信物哦！

除了织带外，凤凰装里还有不少手工编织的精美服饰品。如丽水和平阳的腰巾（图4-10、图4-11），两端就被聪慧的畲家女编织为细密精致的网罗和流苏（图4-12），增加视觉透视感。

These exquisite webbing ties are often used as tokens of love by She girls.

In addition to the webbing ties, there are many hand-woven exquisite accessories in phoenix costumes. Let's take a look at waistbands from Lishui and Pingyang (Fig. 4-10, Fig. 4-11). Intelligent She women weave the edge of the waistbands into fine and delicate nets and tassels (Fig. 4-12), making it more appealing.

图4-10 近代丽水畲族丝腰带[1]

Fig. 4-10 Modern silk waistband in Lishui She ethnic group

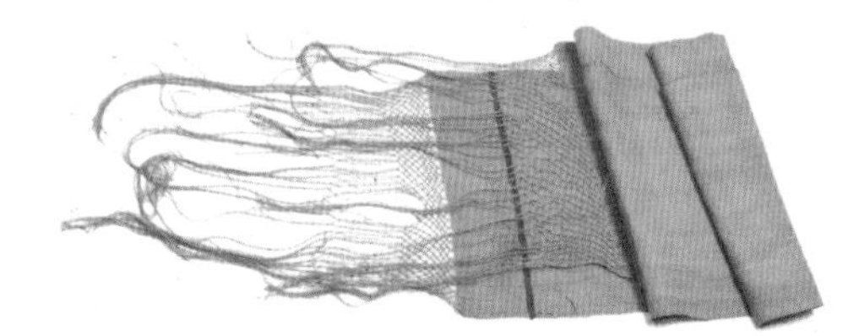

图4-11 近代平阳畲族丝腰带[2]

Fig. 4-11 Modern silk waistband in Pingyang She ethnic group

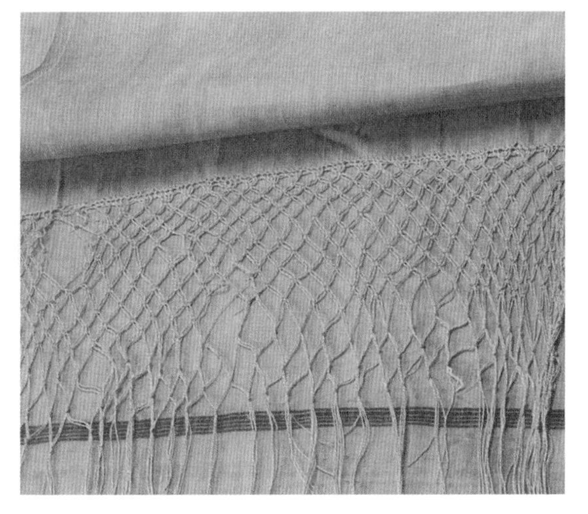

图4-12 近代平阳畲族丝腰带细节

Fig. 4-12 Details of modern silk waistband in Pingyang She ethnic group

[1] 图片来源：征集于丽水地区碧湖沙坑，1959年收藏于浙江省博物馆，长417厘米、宽19厘米，靛青色，近两端处各装饰三条白色弦纹。两端有6厘米网状丝织花边和9厘米长流苏。

Source: collected in the Shakeng Village, Bihu Town, Lishui City and stored in Zhejiang Provincial Museum since 1959. It is 417 cm long and 19 cm wide. The color is indigo. It is decorated with three white chord marks at each end. Both ends have 6 cm mesh silk lace and a 9 cm long tassel.

[2] 图片来源：浙江省博物馆，1959年征集于平阳县桥汀公社乌岩内村。

Source: collected in Wuyannei Village, Qiaoting Commune, Pingyang County in 1959 and stored in Zhejiang Provincial Museum.

图4-13所示钱袋通过管状组织到双层组织的变化，制造出口袋，反映了畲家人在织造技术和空间思维方面的聪明才智。

As shown in Fig. 4-13, the money bag evolves from hollow weave to double weave, thus pockets are created. Such structure reflects the She people's intelligence in weaving and spatial thinking.

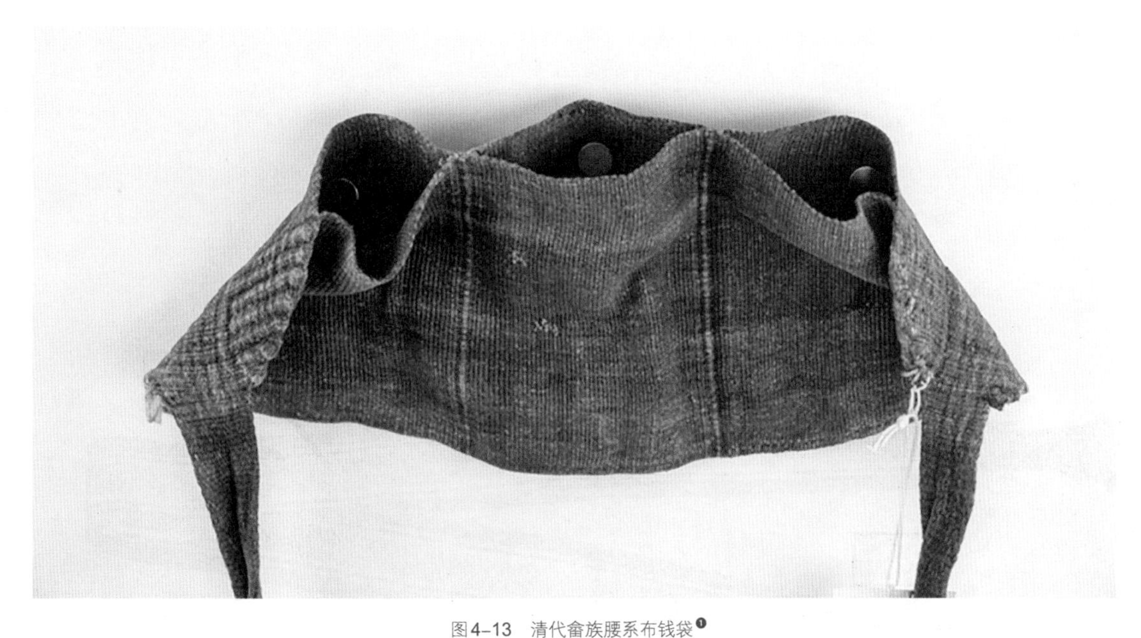

图4-13 清代畲族腰系布钱袋[1]

Fig. 4-13 The She waist money bag in Qing Dynasty

[1] 图片来源：2002年收藏于浙江省博物馆。棉质，长55厘米，宽20厘米。
Source: collected in Zhejiang Provincial Museum in 2002. Made in Cotton, 55 cm long and 20 cm wide.

练习题 | Exercise

你能说出以下纹样的含义吗？

Can you tell the meaning of the following patterns?

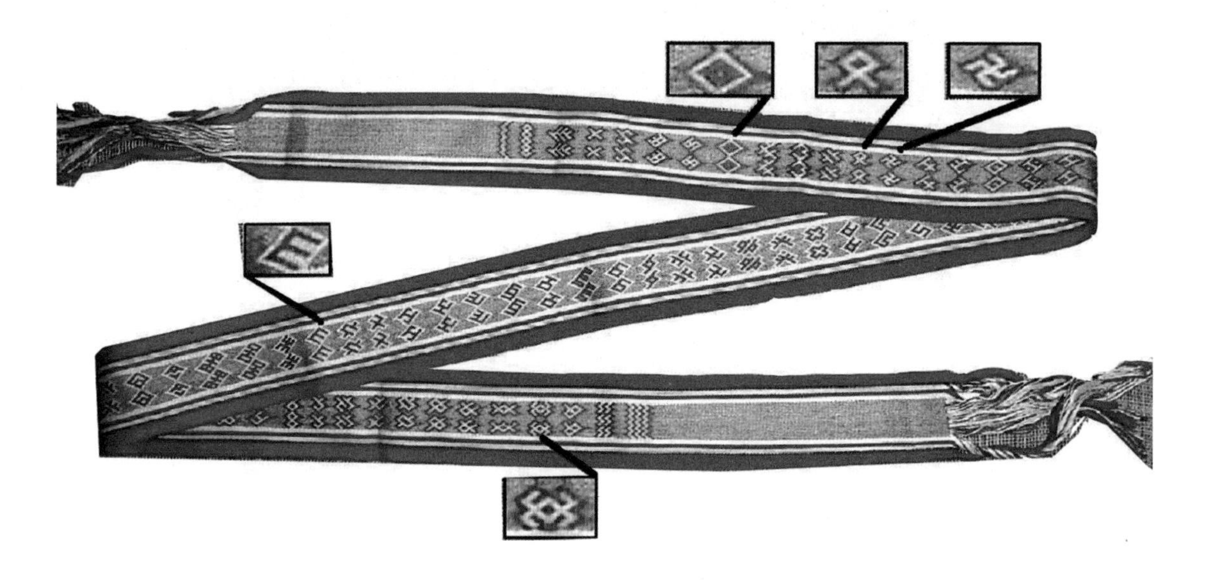

二、刺绣工艺
2 Embroidery

除了“活化石”织带，我最喜欢的就是凤凰装上精致传神的绣花啦！下面这件凤翎式凤凰装的前襟，勾雷形的边框内错落有致地绣满行龙、飞凤、玉兔、瓜果和各类花草，最妙的是在其中还绣着一个双眉倒竖、手舞白练的男子，正在诱钓一只三脚金蟾。这不就是民间广为流传的“刘海戏金蟾”的故事嘛（图4–14）！

Besides the "living fossil" webbing ties, my favorite goes the exquisite embroidery on the costumes. Taking a look at the front of this phoenix costume in phoenix feather style, embroidery of dragons, flying phoenixes, jade rabbits, melons, fruits and flowers are vividly exhibited. The most interesting is an embroidered man waving the white silk with his eyebrows upturned, as he is luring a three-leg golden toad. A-ha, this is exactly the popular folk "Liu Hai plays with golden toad"(Fig. 4-14).

图4–14 温州苍南县凤阳畲服前襟绣花——刘海戏金蟾[1]

Fig. 4-14 Liu Hai plays with golden toad—Embroideries in the front of the garment in Fengyang, Cangnan, Wenzhou

[1] 图片来源：笔者于2004年5月4日摄于浙江省温州市苍南县凤阳畲族乡。
Source: taken in Fengyang She township, Cangnan county, Wenzhou city, Zhejiang province on May 4, 2004.

如图4-15所示围裙，一尺见方的空间里竟绣了40个姿态各异、栩栩如生的人物！而且图案布局井井有条、虚实得当，兼具秩序和韵律感。

Fig. 4-15 shows aprons of Xiapu.Within a foot square, 40 people are vividly embroidered with different postures. How amazing it is! Besides, the layout is in proper order, full of sense of rhythm.

图4-15 霞浦围裙[1]

Fig. 4-15 Aprons of Xiapu

如图4-16~图4-23所示绣花鞋，配色或鲜艳夺目或清雅秀丽，图案或小巧别致或生动形象，每一双都堪称艺术品！

Fig. 4-16 to Fig. 4-23 represent embroidered shows. Be the color light or deep, each of them is artistic. As to their patterns, some are cute and chic, while some are vivid.

[1] 图片来源：笔者于2004年5月4日摄于浙江省温州市苍南县凤阳畲族乡。
Source: taken in Fengyang She township, Cangnan county, Wenzhou city, Zhejiang province on May 4, 2004.

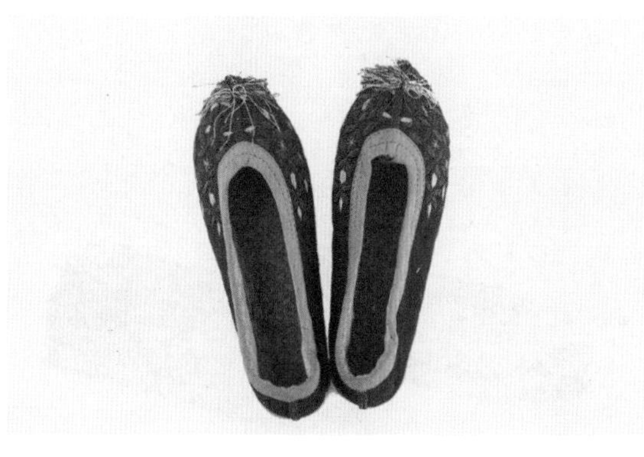

图4-16 雄冠式畲族绣花鞋[1]

Fig. 4-16 The She embroidered shoes in male crown style

图4-17 近代雌冠式畲族绣花鞋[2]

Fig. 4-17 Modern the She embroidered shoes in female crown style

图4-18 近代雌冠式出嫁绣花鞋[3]

Fig. 4-18 Modern embroidered wedding shoes for women in female crown style

---

[1] 图片来源：征集于景宁郑坑吴村雷三妹，2008年收藏于浙江省博物馆，清末民初制作，鞋长25厘米，宽9厘米，高5厘米。

Source: collected from Lei San-mei, Wu Village, Zhengkeng, Jingning and stored in Zhejiang Provincial Museum in 2008. It was made in the late Qing Dynasty and the early Republic of China. The shoes are 25 cm long, 9 cm wide and 5 cm high.

[2] 图片来源：征集于丽水老竹镇占湾村雷珠进处，2001年收藏于浙江省博物馆，鞋尺寸为25 cm×8 cm×7.5 cm。布质，圆头形，千层底。青黑色鞋面，分为左右两侧，由鞋头正中拼缀而成型。

Source: collected from Lei Zhujin in Zhanwan Village, Laozhu Town, Lishui, and stored in Zhejiang Provincial Museum since 2001. The size of the shoes is 25 cm×8 cm×7.5 cm. Made of cloth; round head shape, multiple-layered bottom. It has a blue and black vamp, which is divided into left and right sides, from the middle of the toe patching and forming.

[3] 图片来源：征集于云和县岩下村。1997年收藏于浙江省博物馆，鞋尺寸为23 cm×7.5 cm×8 cm。

Source: collected in Yanxia Village, Yunhe County and stored in Zhejiang Provincial Museum in 1997. The size of the shoes is 23 cm×7.5 cm×8 cm.

图4-19 近代瑞安畲族绣花鞋[1]

Fig. 4-19 Modern embroidered shoes in Rui'an

图4-20 近代平阳畲族绣花鞋[2]

Fig. 4-20 Modern embroidered shoes in Pingyang

（a）

（b）

图4-21 近代泰顺畲族绣花鞋[3]

Fig. 4-21 Modern embroidered shoes in Taishun

---

❶ 图片来源：浙江省博物馆，1959年征集于瑞安县上坎村。

Source: collected in Shangkan Village, Rui'an County in 1959 and stored in Zhejiang Provincial Museum.

❷ 图片来源：浙江省博物馆，1959年征集于平阳县桥汀公社大山村。

Source: collected in Dashan Village, Qiaoting Commune, Pingyang County in 1959 and stored in Zhejiang Provincial Museum.

❸ 图片来源：（a）：浙江省博物馆，1959年征集于泰顺县仕阳公社桥底村。（b）：浙江省博物馆，1959年征集于泰顺雅阳公社下民村。

Source: (a): collected in Qiaodi Village, Shiyang Commune, Taishun County in 1959 and stored in Zhejiang Provincial Museum. (b): Collected in Xiamin Village, Yayang Commune, Taishun in 1959 and stored in Zhejiang Provincial Museum.

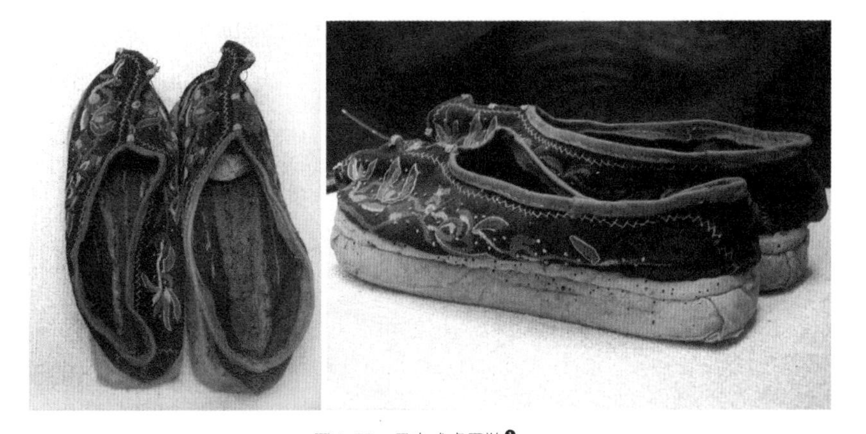

图4–22 凤中式虎牙鞋[1]

Fig. 4-22 Tiger teeth shoes in phoenix trunk style

图4–23 凤头式单鼻虎牙鞋[2]

Fig. 4-23 Tiger teeth shoes in phoenix head style

[1] 图片来源：2018年1月24日摄于厦门大学人类学博物馆。
Source: taken at the Anthropology Museum of Xiamen University on Jan. 24, 2018.

[2] 图片来源：2017年8月21日摄于福建省宁德市中华畲族宫。
Source: taken at the She Minority Palace of China in Ningde, Fujian Province on Aug. 21, 2017.

看了这些畲家刺绣，大家可能注意到凤凰装的绣花主要分布在前襟、领口、袖口、围裙和布鞋上。因为这些部位是日常生活中衣服最容易破损的地方。因此，最早的时候，畲家人用花线密密地穿刺包封这些部位的面料，为的是使之更厚实、牢固，经得起撕扯和磨损。慢慢地，大家开始注意将这些花线安排得错落有致，渐渐形成了花鸟山水和人物的图案。后来，这些刺绣图案还表现出祈求丰收、祝福吉祥的寓意和戏曲诗文、神话传说中的故事场景来（图4–24）。

After seeing these She embroideries, you may notice that the embroidery of the phoenix costumes is mainly distributed on the front, neckline, cuff, apron and shoes, because these are also the areas where clothes are most likely to wear out in daily life. Therefore, the She people threaded needles on these parts in order to make them more thick and firm to withstand the test of time. Later on, these threads are arranged in good order and well-proportioned, gradually forming patterns of flowers, birds, landscapes or figures. They further evolve into patterns that signal a good harvest and blessing, as well as the story scenes in operas, poems and articles, myths and legends (Fig. 4-24).

图4–24　凤凰装上的绣花[1]

Fig. 4-24　Embroidery on phoenix costumes

[1] 图片来源：汤瑛女士于2010年提供。
Source: courtesy of Ms. Tang Ying in 2010.

最后，告诉大家关于畲家刺绣还有一个有趣的风俗，那就是过去专业绣花师傅的都是男性哦！大家尊称他们为“男绣师傅”（图4–25）。

Here I would like to share one more interesting thing about the She embroidery: professionals on this area in the past were all men. We called them "embroider male master" with respect (Fig. 4-25).

图4–25 畲族传统服饰国家级文化遗产传承人蓝曲钗[1]

Fig. 4-25 Lan Quchai, the national cultural heritage inheritor of traditional She costume

[1] 图片来源：2010年4月19日摄于福建省罗源县竹里村。
Source: taken in Zhuli Village, Luoyuan County, Fujian Province on April 19, 2010.

练习题 | Exercise

请为以下畲族服饰填上你喜欢的色彩吧！

Please color in pictures below for the She costumes.

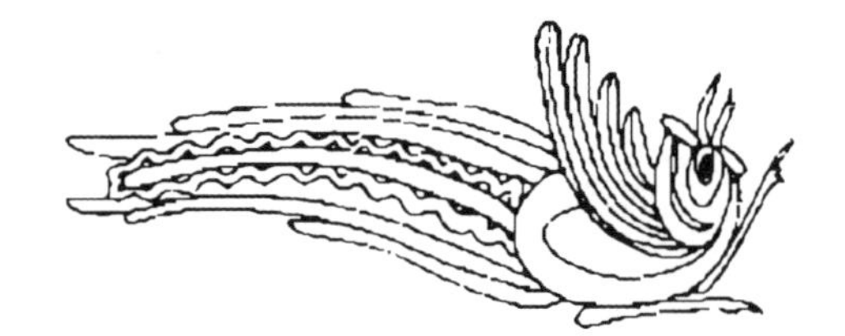

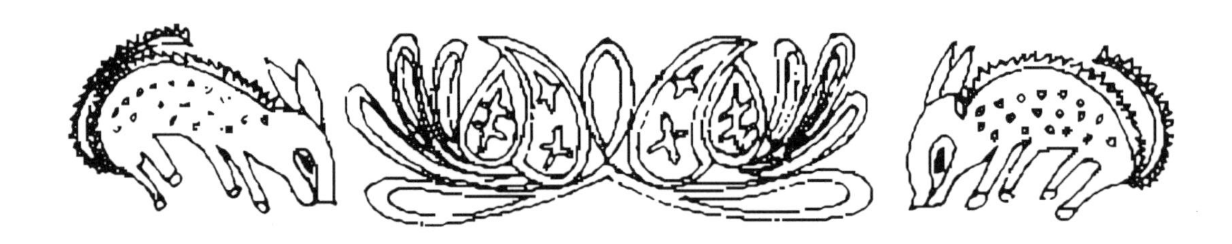

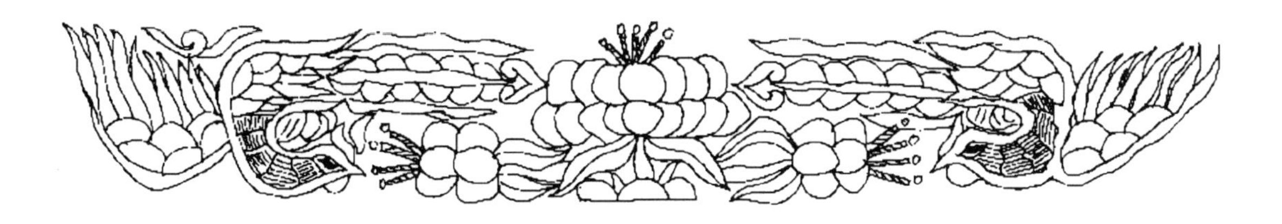

三、染色工艺

3 Dyeing

畲家有句谚语："吃咸腌，穿青蓝。"正像前面跟大家介绍的一样，几乎各个地区的传统凤凰装都以青色（深蓝近黑）或蓝色为主色调。这主要是因为我们畲家人特别擅长染蓝布！我们在古代曾经被称为"菁客"，这"菁"就是蓝草，是用来制作染料"蓝靛"的原料，而用蓝靛就可以染出漂亮的蓝色和黑色。我们常说的"青出于蓝而胜于蓝"里的"青"就是指蓝靛。畲家人从明代开始就在山上搭棚种蓝靛，熊人霖著《南荣集》记载崇祯年间闽西南"汀之菁民，刀耕火耨，艺兰为生"（"兰"同"蓝"）；《兴化县志》也记载着闽中莆仙畲民"彼汀漳流徙，插菁为活"。据明代黄仲昭《人间通志》卷四一记载，"菁客"所产蓝靛品质极佳，其染色"为天下最"！

A She saying goes that "eat preserved and salted food, wear indigo blue". As introduced before, the main color of the traditional phoenix costumes is indigo blue (dark blue nearly black) or blue as the She are very good at dyeing blue. We used to be called "Jing Ke" (indigo producer) in ancient times. Jing refers to the indigo plant, which is used to dye the color of indigo, a beautiful color mixed with blue and black. In the popular proverb "Indigo comes from blue and is better than the latter. The She people had built shelters to grow indigo in the mountains since the Ming Dynasty (1368-1644). In *Nanrong Collection*, a book authored by Xiong Renlin, recorded that during the Chongzhen period, southwestern Fujian witnessed the She people making a living by planting indigo plants and processing them. *Xinghua County Record* also recorded the She people in Putian, Fujian migrating between Ting (today's Changting, west of Fujian) and Zhang (today's Zhangzhou, southeast of Fujian) and making a life by planting indigo". According to *Renjian Tongzhi*, authored by Huang Zhongzhao (Ming Dynasty), the indigo produced by "Jing Ke" is of excellent quality, and its color is "the best of the best"!

民国以前，我们都是用植物染料自制自染凤凰装，民国以后逐渐变成到集镇上找专门染色的作坊漂染。再往后，挑担郎的出现使我们可以直接买到青、蓝、紫、红等色的染色粉，染线过程就更简便了，可供选择的颜色也更多，服饰的用色就更为丰富了（图4–26）。

Before Republic of China, phoenix costumes were homemade. After that, we gradually turned to dye workshops in towns for bleaching and dyeing. Later on, peddlers, made things easier as we could buy dye powder in green, blue, purple, red and other colors. The dyeing process was simplified and we got more colors to choose from, contributing to more colorful phoenix costumes(Fig. 4-26).

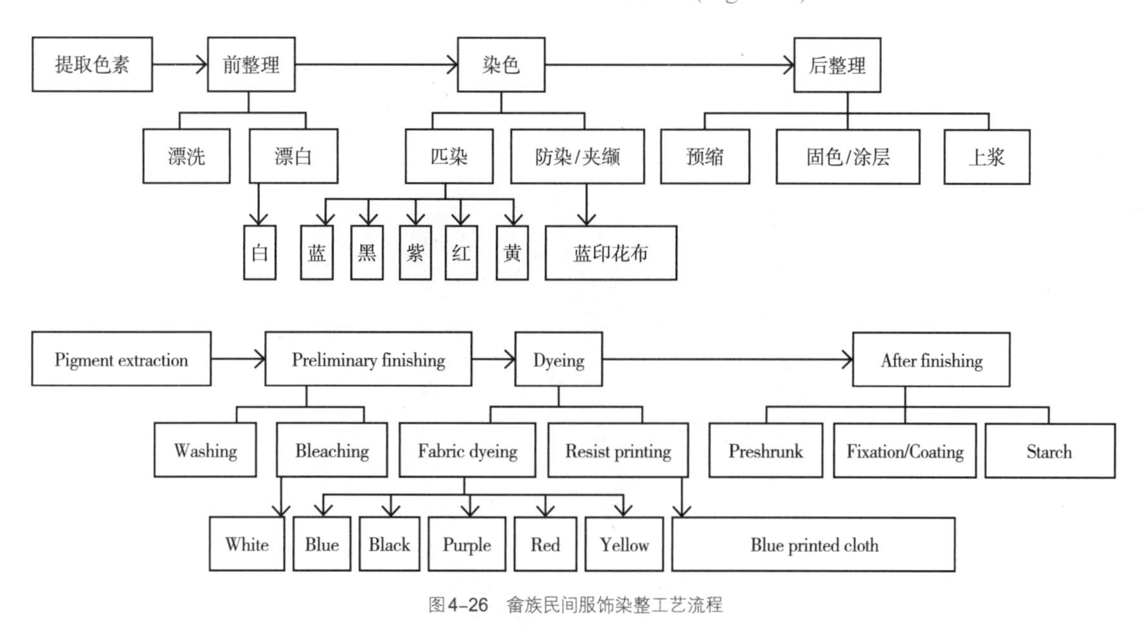

图4–26　畲族民间服饰染整工艺流程
Fig. 4-26　Dyeing and finishing process of She costumes

畲家人的染色技艺主要包括蓝靛色素提取和蓝色染色工艺。蓝染工艺又分为蓝色匹染工艺和蓝印花布染色工艺。

The dyeing skills of the She people, mainly including indigo pigment extraction and blue dyeing. Blue dyeing technology is divided into match dyeing and printing technique.

1. 蓝靛色素提取

1. Extraction of indigo pigment

清代畲民制蓝原料包括蓼蓝、染蓝和大蓝。[1]近代畲族主要采用蓼蓝来染制青蓝色面料，包括棉麻类服装面料和蚕丝制织带线（图4–27）。

The raw materials of producing indigo by the She people in the Qing Dynasty include Liaolan (Polygonum tinctorium ait), Ranlan and Dalan[1]. In modern times, the She people mainly use Liaolan to dye blue fabrics, including cotton, hemp and silk (Fig. 4-27).

图4–27 蓝草之蓼蓝[2]

Fig. 4-27 The Polygonum tinctorium Ait of bluegrass

---

[1] 据清《闽产录异》："闽诸郡多种蓝，曰蓼蓝，曰染蓝，曰大蓝，皆可作淀。"

According to the book *Records of Fujian Productions* in Qing Dynasty: "There are three kinds of bluegrass in the counties of Fujian, namely Liaolan, Ranlan and Dalan. All of which can be used as the source of indigo."

[2] 图片来源：2004年8月6日笔者摄于贵州省台江县。

Source: taken by the author in Taijiang County, Guizhou Province on August 6, 2004.

根据顾炎武《天下郡国利病书》的记载，近代畲族制作青靛的方法是每年霜降后割蓝，“凌举桶中，再越宿乃出其枝梗，纳灰疾搅之，泡涌微白，文之浙青，泡尽靛花与灰俱降，及澄蓄之，而泻出其水，则靛可滤而染”。即将蓝草收割后放至桶中浸泡，过一夜后取出其中的枝和梗，再向桶中加入石灰快速搅拌，一开始桶内液体冒出的泡沫微微泛白，过段时间后就变为深蓝色，当泡沫消失，让靛蓝染料和石灰一起沉淀，倒掉水后，滤出的靛蓝就可以用于染色了（图4–28）。

According to Gu Yanwu's book *Political and Economic Conditions throughout Ming Empire*, indigo making process at that time included the following steps: cut the indigo plant after the Frost's Descent (the last solar term of autumn) every year; put the indigo plant after harvest to soak in the barrel; take out the branch and stem after a night; stir in lime quickly in the barrel. At first, the bubbles of the liquid in the barrel are slightly white, later they become dark blue. When the bubbles disappeare, indigo dye and lime precipitate together. After pouring out the water, the indigo filtered out can be used for dyeing (Fig. 4-28).

图4–28 蓝靛色素提取[1]

Fig. 4-28 Extraction of indigo pigment

[1] 根据霞浦水门乡大坝村村民蓝建杯口述制作。本节参考《闽东畲族文化全书：服饰卷》。
Oral record from Lan Jianbei, villager from Xiapu Shuimen Township Dam Village. This section refers to *The Complete Book of She Culture in Eastern Fujian: Costume*.

2 蓝色染色工艺[1]

2. Blue dyeing[1]

（1）蓝色匹染工艺

(1) Match dyeing

①染色前先做漂洗准备：先将带浅黄色苎麻坯布浸入高2.1米、直径约2米的木桶中，浸泡约2个小时漂洗杂质，再将坯布取出拧干，挂在横杆上晾干。

②将发酵蓼蓝、土茯苓的浓汁加入染缸调匀，再将布浸入染缸中。加入土茯苓根汁是为了保持颜色鲜艳不褪色，即提高色牢度。将水加至温热，师傅手持两根木棒，来回搅拌，再用木棒夹住布使劲反复拧，目的是使布染得均匀，俗称“拷青”“拷蓝”，又称“染缸”。

③染匀后取出布匹放在微火上烤干。

① Rinsing: soak the pale yellow ramie fabric in a wooden barrel which is about 2.1 meters high and 2 meters in diameter for about 2 hours to wash the impurities. Then take out the fabric, wrest it and hang it on a horizontal pole to dry.

② Dyeing: add the thick juice of the fermented indigo and Tufuling (Glabrous Greenbrier Rhizome) into the dye vat and put the fabric in it. Combine the juice of root of Tufuling to keep color bright and not fade, namely to raise color fastness. Once the water is warmed, the worker stirs the fabric back and forth with two wooden sticks, and then holds it with the sticks and twists it hardly and repeatedly. Thus the fabric can be dyed evenly. The whole dyeing process is commonly known as "Kao Qing" "Kao Lan", also known as the "Ran Gang (vat dyeing)".

③ Once dyeing is finished, the fabric will be dried with slow fire.

---

[1] 根据霞浦水门乡大坝村村民蓝建杯口述制作。本节参考《闽东畲族文化全书：服饰卷》。
Oral record from Lan Jianbei, villager from Xiapu Shuimen Township Dam Village. This section refers to *The Complete Book of She Culture in Eastern Fujian: Costume.*

（2）蓝印花布染色工艺

(2) Printing technique

先制作一块与面料门幅相同的薄木板或油纸板，然后用尖刀镂刻出松海花瓣，紧贴在面料上，用黄豆浆与米浆相混合，用刷子多次刷在漏刻花瓣上，使其凝固，再刷上灰水待干后，亦按靛青程序染。经漂洗晾干后，将灰块刮去则花部位呈现白色（图4–29）。

Printing technique observes the following steps: make a piece of thin board or oil board with the same size of the fabric; use a sharp knife to skip out the flower petals; closely attach the board to the fabric; mix soybean milk with rice milk, and brush the milk repeatedly on the flower petals on the board; once the milk is solidified, brush it with lye. Having done that, following processes of match dyeing. After rinsing and drying, scrape off the ash, then the flower will be white(Fig. 4-29).

图4–29 凤头式盛装蓝印花腰带[1]

Fig. 4-29 The waist tie (printed technique) in phoenix head style

[1] 图片来源：景宁中国畲族博物馆供图。

Source: She Museum of China in Jingning.

练习题 | Exercise

请参照畲族服饰图案画一画吧!

Please draw a picture according to the following pattern of She costume.

四、制银工艺

4 Silvering

银饰是凤凰装中必不可少的配饰。对各地凤凰装来说都是最具标志性的特征的凤冠，必然含有银髻牌等银制部件（图4-30）。除

Silver jewelry is an indispensable accessory for phoenix costumes. For all phoenix crowns, silver parts are the most iconic features of all phoenix costumes(Fig. 4-30). Represented by the silver

图4-30 制银[1]

Fig. 4-30 Producing the silver ornaments

[1] 图片来源：2017年8月22日摄于福建省福安市穆云畲族乡后舍村（银匠村）。

Source: taken in Houshe Village (Silversmith Village), Muyun She ethnic group township, Fu'an city, Fujian province on Aug. 22, 2017.

此之外，从头笄、耳环（图4-31）到手镯（图4-32），再到项链（图4-33）、戒指（图4-34），还有银头花（图4-35）、各式钗钏（图4-36），连童帽上也以银饰装点（图4-37），品类繁多，数不胜数。仅霞浦县1982年由有关部门分配的特需银饰品就达2000余件，折合白银14.25公斤呢！

Jipei, they are must-haves. In addition, many other accessories, like hairpins, earrings (Fig. 4-31), bracelets (Fig. 4-32), necklaces (Fig. 4-33), rings (Fig. 4-34), as well as silver headdress flowers (Fig. 4-35), various hairpins (Fig. 4-36), and even children's hats (Fig. 4-37), are decorated with silver ornaments. There are way many to mention. The silver ornaments for special needs allocated by the relevant government department to Xiapu county in 1982 amounted to more than 2000 pieces, equivalent to 14. 25 kg of silver.

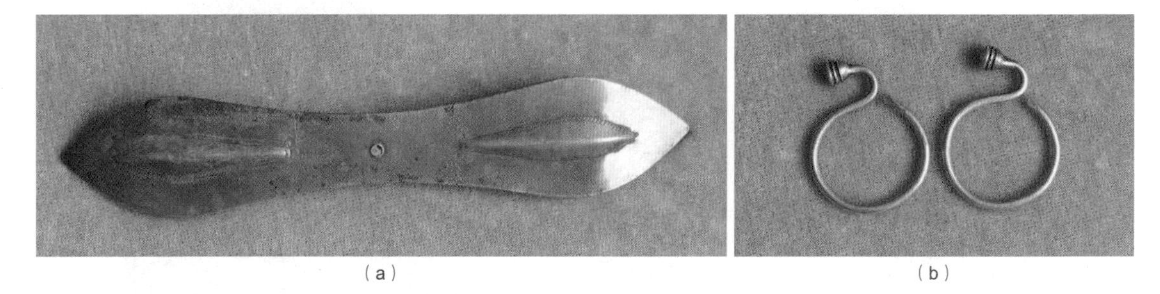

(a) (b)

图4-31 霞浦畲族妇女头笄及少女大耳环[1]

Fig. 4-31 Women's hairpin and girls' earrings in Xiapu

[1] 图片来源：由汤瑛女士于2010年提供。
Source: courtesy of Ms. Tang Ying in 2010.

图4-32 畲族典型手饰之闽东九圈镯❶

Fig. 4-32 Typical She bracelets: a pair of silver bracelets with nine loops (eastern Fujian)

图4-33 畲族典型胸饰之福安花篮式银胸牌

Fig. 4-33 Typical She necklaces: silver in flower basket shape (Fu'an)

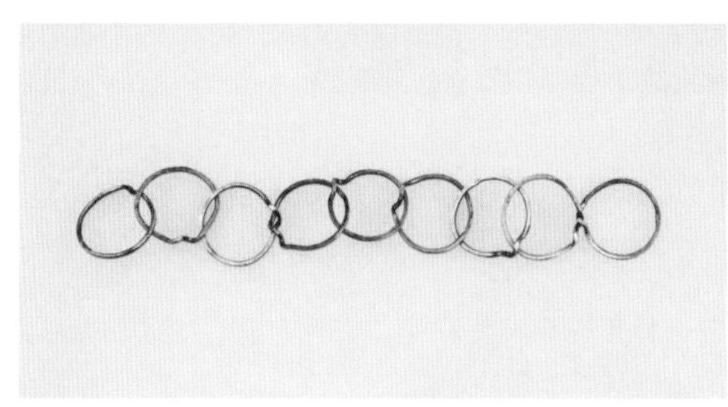

图4-34 畲族典型手饰之浙南九连环戒指❷

Fig. 4-34 Typical She ring: nine linked rings in southern Zhejiang

❶ 图片来源：宁德市“盈盛号”金银饰品有限公司林贤学董事长提供。2010年上海世博会福建馆指定礼品。
Source: provided by Lin Xianxue, Chairman of Ningde "Yingshenghao" Gold and Silver Accessories Co., Ltd. Exclusive gifts of Fujian Pavilion in 2010 Shanghai World Expo.

❷ 图片来源：由景宁畲族博物馆提供。
Source: provided by the She Museum of China in Jingning.

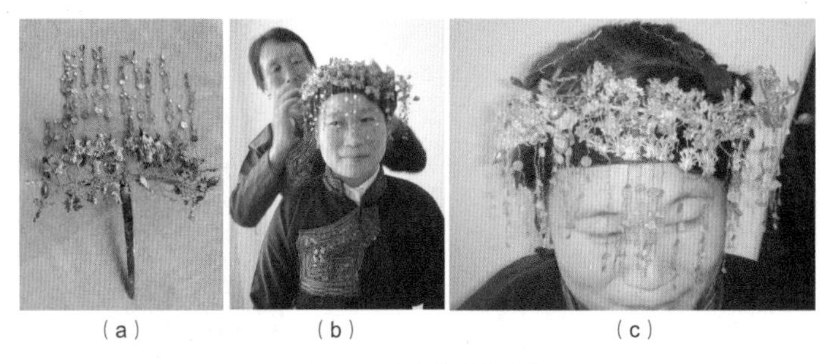

(a) (b) (c)

图4-35　福鼎新娘银头花❶

Fig. 4-35　Silver headdress of the bride in Fuding

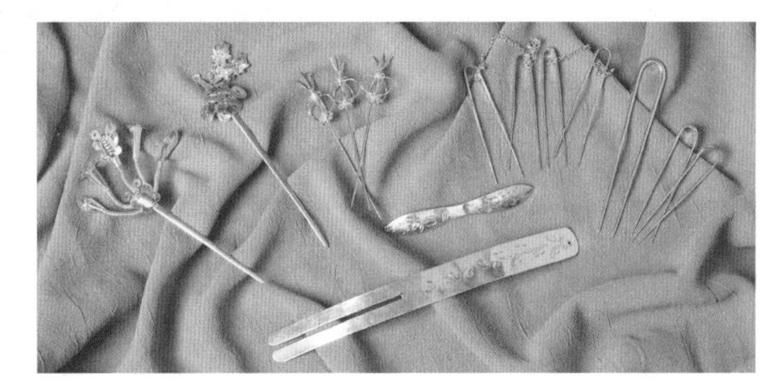

图4-36　部分畲族头饰

Fig. 4-36　She headwears

图4-37　畲族典型帽饰之一——近代浙江丽水地区畲族小孩福帽❷

Fig. 4-37　Typical She hat—modern blessing cap for kids in Lishui of Zhejiang

❶ 图片来源：(a)：由汤瑛女士于2010年提供；(b)、(c)：由福鼎市民族与宗教事务局钟敦畅先生提供。

Source: (a): courtesy of Ms. Tang Ying in 2010. (b),(c): Courtesy of Mr. Chung Dun-cheong from Fuding Bureau of Ethnic and Religious Affairs.

❷ 文物来源：浙江省博物馆。

Source: Zhejiang Provincial Museum.

畲族银饰造型别致，独具匠心。图4-38所示的泰顺地区的畲族戒指戒面为梅花图案，涂景泰蓝釉，戒指背面有“玉成”两字，祝福婚姻幸福美满。图4-39所示戒指戒面凸出成立体的兽首状，且兽首两眼有孔，孔内有两只可以活动的触角，十分有趣，有学者认为是盘瓠图腾崇拜的表现。

The She silver ornaments have a unique shape and originality. As shown in Fig. 4-38, the She ring in Taishun, southern Zhejiang, is decorated with a plum pattern and painted with cloisonne glaze. The word "Yu Cheng" (literally jade and success) is on the back of the rings, wishing a happy marriage. As shown in Fig. 4-39, the ring has a three-dimensional animal head, with a hole in each eye, which contains movable antennae. Some scholars believe such an interesting pattern represents the worship of the Panhu totem.

图4-38 近代泰顺畲族梅花戒指[1]

Fig. 4-38 Modern plum blossom ring of She ethnic group in Taishun

图4-39 近代泰顺畲族兽首戒指[2]

Fig. 4-39 Modern animal head ring of She ethnic group in Taishun

---

[1] 图片来源：浙江省博物馆，1959年征集于浙江省泰顺县司前白岩村。

Source: collected from Siqian Baiyan Village, Taishun County, Zhejiang Province in 1959 and later stored in Zhejiang Provincial Museum.

[2] 图片来源：浙江省博物馆，1959年征集于浙江省泰顺县彭溪公社仓基村。

Source: collected in Cangji Village, Pengxi Commune, Taishun County, Zhejiang Province in 1959 and stored in Zhejiang Provincial Museum.

除装饰和祈福外，畲族银饰也是传统畲族生活中用以含蓄表达身份的一种标识，例如福鼎畲族女性婚前婚后所带银饰有所区别：婚前女子头梳长辫，扎大红绒毛线，戴耳牌，耳牌是畲族耳饰的一种，上部是曲形银钩，下面挂着錾有花草图案的三脚银牌，如图4–40（a）所示；婚后女子梳盘龙髻，戴耳燕，耳燕是畲族耳饰的一种，形如问号，如图4–40（b）所示，髻插银针、银花。有畲族民歌这样唱道："表妹做人眼未斜，一边耳朵会（挂）耳牌，一边耳朵会耳燕，问你几（何）时哭母鞋（唱哭嫁歌）？"

Silver jewelry goes beyond decoration and blessing for the She people. It is also a symbol revealing the marital status of She female. Take women in eastern Fujian as an example. Before marriage, women there wear long braids with bright red threads, and ear tags [Fig. 4-40(a)]. After marriage, they comb a Panlongji (dragon bun), wear ear swallow [Fig. 4-40(b)], and insert silver needles and silver flowers on the bun. Some She folk songs go like this: "My cousin's eyes are not inclined, but one ear wear ear tag, one ear wear ear swallow, so when will you sing the marriage song?"

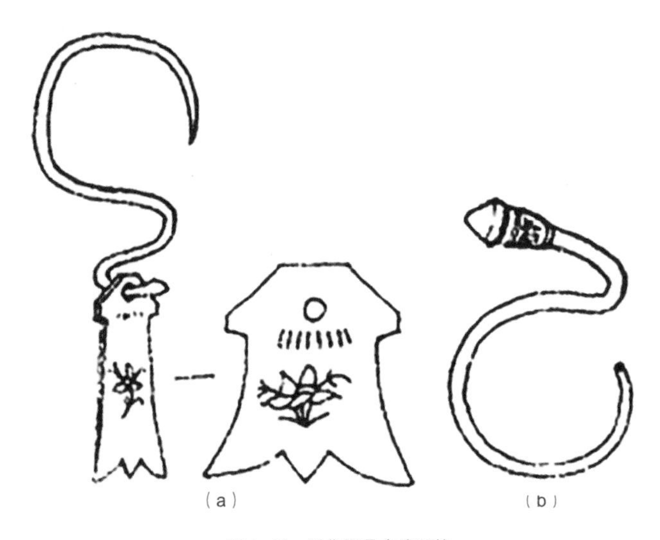

图4–40　近代福鼎畲族耳饰

Fig. 4-40　Modern ear rings of the She minority in Fuding

练习题 | Exercise

请参照以下畲族服饰图案画一画吧！

Please draw a picture according to the following pattern of She costume.

五、裁剪工艺
5 Tailoring

畲家人在设计制作服装时始终贯彻着“物尽其用、精工细作”的原则。绝大部分服饰品都明显地体现出制作者在设计之初就考虑到了如何对材料进行最为充分的利用，下面要说到的绑腿、围裙和包袋的裁剪方法都能做到零损耗用料，给现代环保设计提供了借鉴。

绑腿在畲语[1]里称为“脚绑”，是各地区畲族普遍使用的服饰品，用于在山间林地保护小腿不被荆棘利草割伤。汉族常用的绑腿尺寸可变化，没有严格标准，单个绑腿一般为一条长布带，宽度约10厘米或更宽，长度为1.5~2米。畲族常用的绑腿如图4-41所示，单个绑腿为一块长方形面料两头向中心折叠后缝制而成，折

The She people always abide by the principle of "making the best use of everything and delivering fine works" when designing and making clothes. This can be exemplified in most of the garments, whose design fully exhibits designers' delivery of the principle. The tailoring methods of puttees, aprons and bags mentioned below cause zero loss of materials, which provides references for modern green design.

Puttee is called[1] "Jiaobang" (foot binding) in the She language. It is an accessory widely used by the She people in various regions as it can protect the leg from being hurt by bramble and grass in the forest and mountains. The size of the puttee commonly used by the Han people varies from place to place. There is no universal standard. A single puttee is about 10 cm wide, and 1.5 to 2 m long. The puttee commonly used by the She people is shown in Fig. 4-41. A single She leg binding is made of a piece of rectangular fabric with both ends folded toward the center. When folding, one

[1] 畲语是畲族的本民族语言，属于苗瑶语苗语支，目前仅有中国广东省境内的极少数畲族使用。畲语没有文字。
She language is a Hmong-Mien language used in Guangdong province in southern China. It is a spoken language instead of a written one.

叠时一端对折，另一端沿45° 对角线折叠。绑腿用料为宽度约25厘米、长度约50厘米的长方形，比汉族绑腿用料节省了许多。通过对面料进行分割制作，不产生任何边角料，从而达到了用料零损耗。

如图4–42所示为当代樟坪乡当地围裙，用料1米（3尺），完全没有边角料剩余，利用率可达100%。

end is folded in half and the other end is folded along the diagonal of 45°. They are made of rectangular material with a width of about 25 cm and a length of about 50 cm. Thus it is concluded that the She's puttee save a lot of materials than that of Han as the former arguably achieve zero loss.

As shown in Fig. 4-42, the local apron of Zhangping Town is made of one meter-long material, with no leftover. In a word, zero loss.

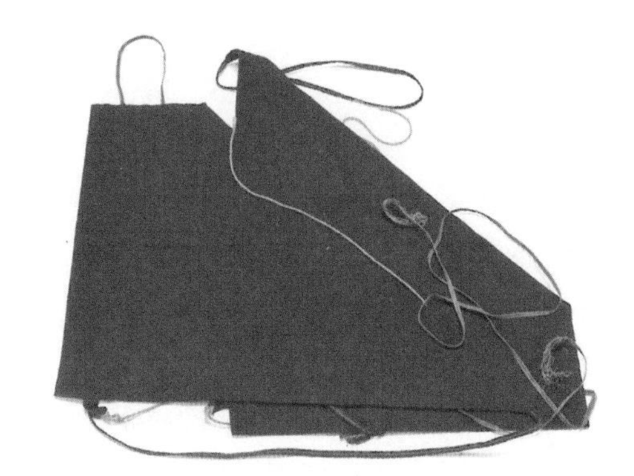

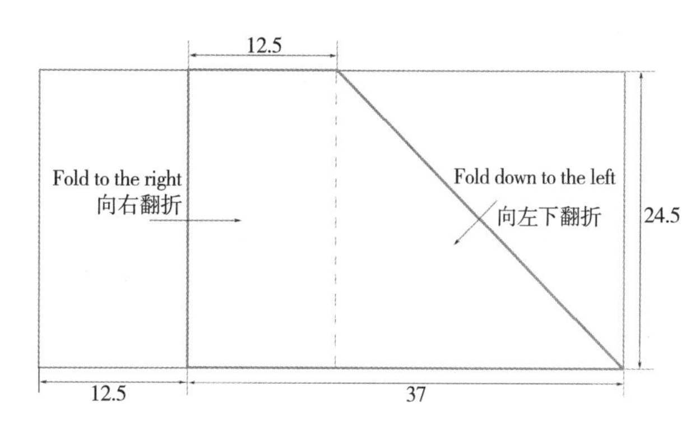

图4–41 清代丽水畲族绑腿及其裁剪图（厘米）[1]

Fig. 4-41 The She puttee in Lishui in Qing Dynasty and its tailing pattern (cm)

[1] 图片来源：征集于丽水市云和县垟岗，1959年收藏于浙江省博物馆，长24. 5厘米、宽37厘米，带长83厘米，面料藏青色，外观呈直角梯形，直角边两端各连接一条绑带。

Source: collected in Sanyang Gang, Yunhe County, Lishui City and stored in Zhejiang Provincial Museum in 1959. It is 24.5 cm long, 37 cm wide and 83 cm long. The fabric is navy blue. The appearance is a right angle trapezoid, both ends of the right angle edge are connected with a webbing tie.

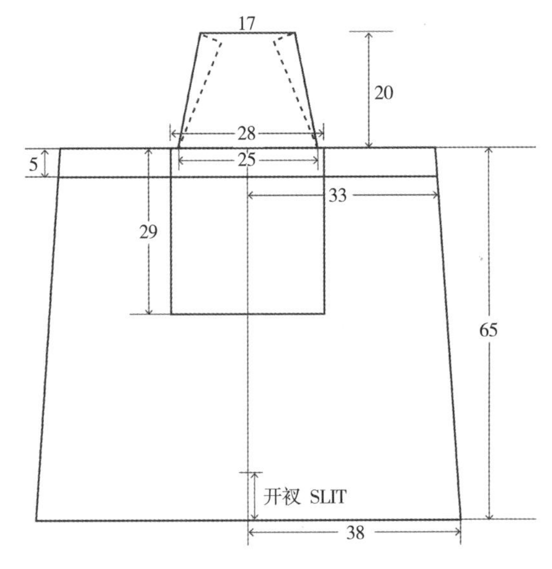

图4-42 贵溪市樟坪畲族乡当地围裙（厘米）❶

Fig. 4-42 Apron in Zhangpin She Ethnic Township in Guixi City (cm)

相对于服装，包或袋的设计受人体约束较少，设计更加自由。畲族各地传统包袋造型各异，但都不约而同地采取了零损耗的用料方式。如图4-43所示，杭州桐庐莪山和温州文成的畲族都流行一种结构十分巧妙的提袋，是用两块矩形的面料斜拼而成，留出的两个角打结或系绳，即为便于拎提的手柄。

Compared with costumes, the design of the bags is less constrained by the human body. The traditional bags of the She people, though in different shapes, all observe zero loss. As shown in Fig. 4-43, the carrying bag of a very ingenious structure is popular among the She people in Tonglu, Hangzhou and Wencheng, Wenzhou. The bag is made of two rectangular fabrics pieced at an oblique angle, and the two corners left are the handles that are easy to carry after tying knots or tying ropes.

❶ 图片来源：笔者2012年8月15日摄于江西贵溪市樟坪畲族乡。

Source: shot by the author in She ethnic group Township in Zhangping, Guixi City, Jiangxi Province on Aug. 15, 2012.

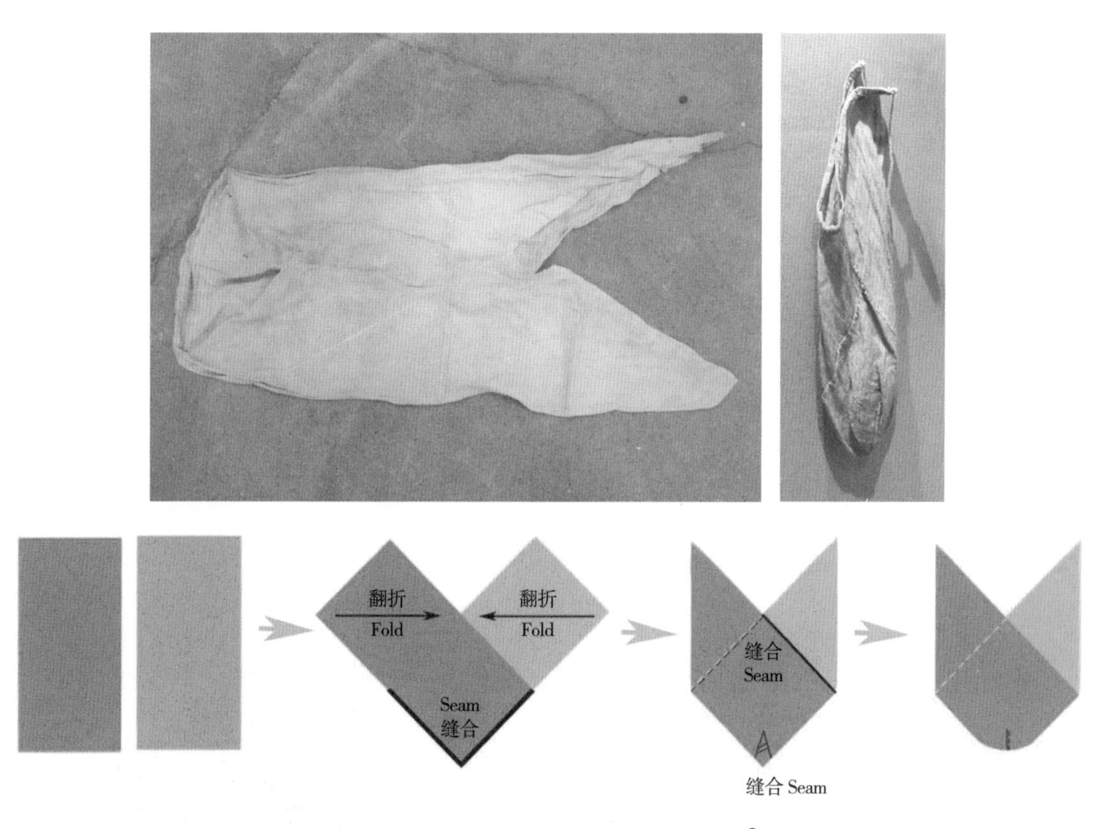

图4-43　桐庐莪山和温州文成的畲族提袋及其结构[1]

Fig. 4-43　Carrying bags popular in Tonglu and Wenzhou

除了节约的意识，畲家裁剪技艺还显露出在美观性和舒适性等方面的巧思。例如，畲族肚兜的设计就很巧妙，如图4-44所示，将口袋设计为贴着肚兜下端的U形，口袋上端由两个扇形折枝花适合纹样固定，整个肚兜以黑色贴

In addition to the awareness of saving, the tailoring skills of the She people also show clever thinking in the aspects of beauty and comfort. For example, the design of the She's Chinese bellyband is very ingenious. As shown in Fig. 4-44, the pocket is designed to be U-shaped at

[1] 图片来源：分别由桐庐莪山畲族乡乡政府和温州文成博物馆畲族馆提供。

Source: provided by the She ethnic group Township Government of Eshan, Tonglu and the She ethnic group Museum of Wencheng Museum, Wenzhou respectively.

边，白色作底，上部两枚红白相间贴布绣，清雅而生动。如图4–45所示的肚兜沉静朴素，却也别有新意，只在中间用紫红色线绣出两只小三角固定袋口，将中间的口袋“隐形”起来。

the lower end of the Chinese bellyband, and the upper end is fixed by two fan-shaped flowers. The whole Chinese bellyband is trimmed in black with white as the bottom, and the upper part is composed of two red and white appliques, which are elegant and lively. As shown in Fig. 4-45, the Chinese bellyband is simple but pragmatic. Two small triangles are embroidered in the middle to fix the mouth of the bag, and the middle pocket is "invisible".

图4–44　民国畲族肚兜❶

Fig. 4-44　The Chinese bellyband of the She ethnic group during Republic of China

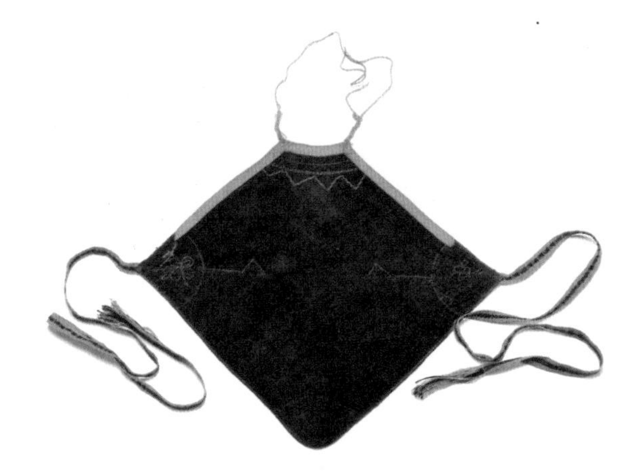

图4–45　畲族肚兜❷

Fig. 4-45　The Chinese bellyband of the She ethnic group

❶ 图片来源：征集于平阳县桥汀公社乌岩内村，1959年收藏于浙江省博物馆。长62厘米，宽48厘米。
Source: collected from Wuyannei Village, Qiaoting Commune, Pingyang County, and stored in Zhejiang Provincial Museum in 1959. It is 62 centimeters long and 48 centimeters wide.

❷ 图片来源：征集于丽水地区，1959年收藏于浙江省博物馆。长40厘米，宽35厘米。
Source: collected in Lishui area and stored in Zhejiang Provincial Museum in 1959. It is 40 cm long and 35 cm wide.

同样用于系腰，图4–46所示腰带的设计初衷不只是为了审美，更多是舒适性和功能性，它巧妙地用了两条布条呈螺旋形缝制，使沿腰带方向的丝缕方向正好是面料45° 斜丝，充分利用面料的延展性，加大了腰带的弹性。

The waist tie, as its name indicates, is to be tied on the waist. As shown in Fig. 4-46, it is good-looking, comfortable and pragmatic. It cleverly uses two cloth strips to sew in a spiral shape, so that the angle between the direction of the fabric grain line and the direction of the belt is exactly 45°, making full use of the ductility of the fabric and increasing the elasticity of the tie.

图4–46 畲族腰带[1]

Fig. 4-46 Waist tie of the She minority

[1] 图片来源：2003年收藏于浙江省博物馆。棉质，长280厘米，宽9厘米。

Source: collected in Zhejiang Provincial Museum in 2003. Made of cotton, 280 cm long and 9 cm wide.

练习题 | Exercise

请为以下畲族服饰填上你喜欢的色彩吧！

Please color the picture below for the She costumes!

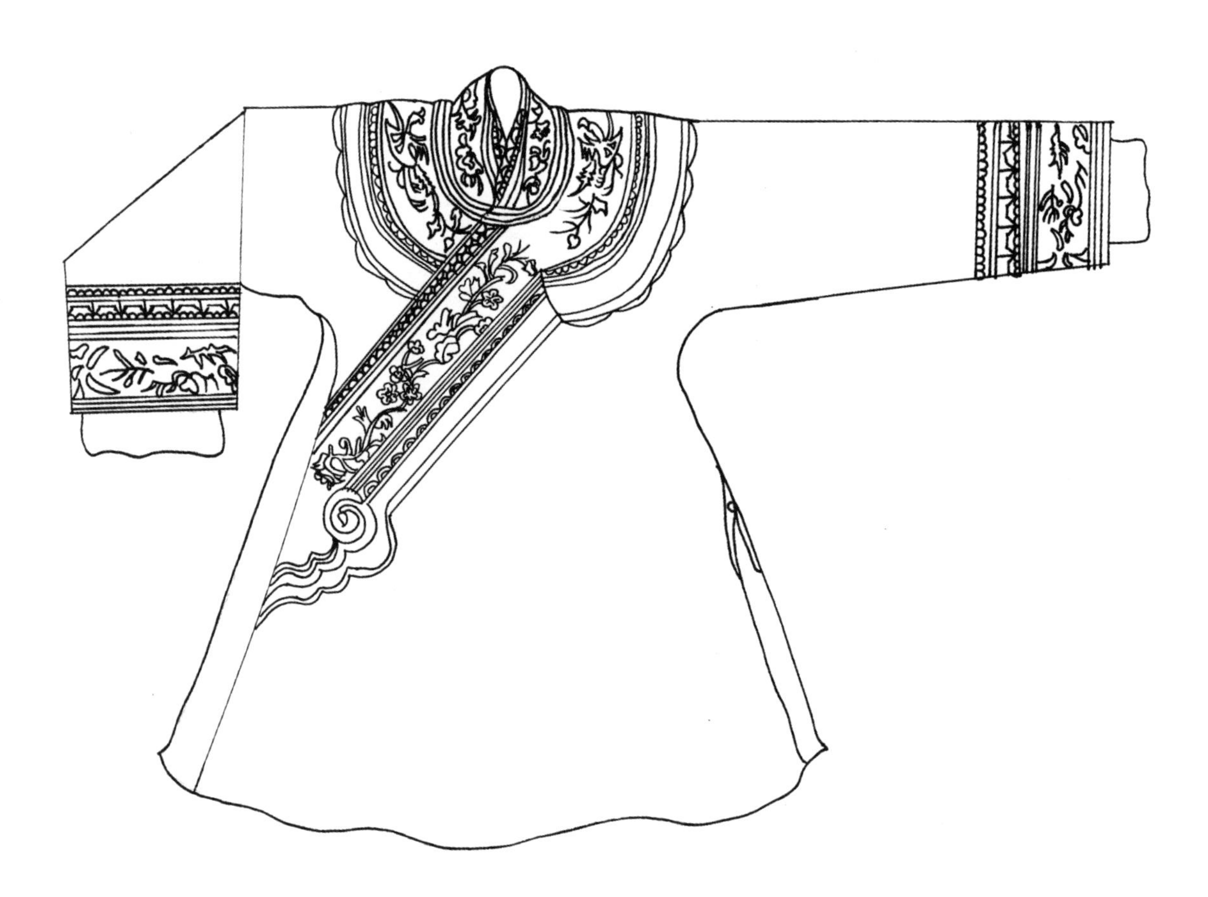

第五章
Chapter
5
畲族服饰变迁脉络
Development Patterns of She Costumes

畲族人民历经千年迁徙，分散在华东地区和贵州的广大山区，各地畲族服饰各有特色的同时也存在内在联系，呈现纷繁复杂的局面。1985年，潘宏立先生依据各类福建畲族服饰结构的相似性，推断其中的密切联系反映了各类型服饰的渊源关系，进而提出“畲族服饰的差异是地方性变异”。潘先生认为从畲族文化历史发展的总体过程来看，畲族服饰是从统一发展到多样，逐渐形成地方差异的，并指出福建畲族服饰差异是迁徙散居后基于地理、文化隔离而产生的辐射变异。2017年陈敬玉在《民族迁徙影响下的浙闽畲族服饰传承脉络》一文中将浙闽畲族服饰归纳为5种典型分支样式，与2条由闽入浙的历史迁徙路线进行对照，发现其贯穿着该5种服饰，且具有一定的脉络相承关系，故指出畲族服饰脉络相承关系与民族迁徙路径之间具有较为一致的对应性：“畲族服饰由罗源式为起点至景宁式为终点，途经福安、霞浦、福鼎和泰顺，存在一脉相承的连贯性。”

The She are scattered in the vast mountainous areas of East China and Guizhou now after thousand-year migration. The She costumes in those places keep their unique characteristics while sharing something in common. In 1985, Mr. Pan Hongli, based on the similarity of various She costumes in Fujian, inferred that the close connection of the costumes reflected the relationship among them, and thus proposed that "the differences of She costumes are local variations". Mr. Pan believed that from the overall development of the She history and culture, She costumes have developed from one to many, and gradually formed local characteristics. He added that She costumes in Fujian were variants based on the geographical and cultural realities after a long-time migration. The article *The Inheritance of Zhejiang and Fujian Costumes Under the Influence of Ethnic Migration*, published by Mr. Chen Jingyu in 2017, summarized five styles of She costumes in Zhejiang and Fujian after comparing the two migration routes from Fujian to Zhejiang. He found that places along the routes wear the five types of costumes at different times and share certain similarities. Thus he concluded that the styles of these costumes had a certain relationship with the migration routes: "She costumes, starts from Luoyuan style and ends at Jingning style and styles between of the two include Fu'an, Xiapu, Fuding and Taishun style. Each of them exhibits legacy from the last station."

在前人研究的基础上将研究范围和类型进一步增大，将华东地区畲族服饰分门别类进行梳理，可以发现其头饰、耳饰、女鞋、襟角造型均存在明显的内在渐变联系。

畲族盛装头饰造型变化脉络如图5-1所示，以罗源式为源头，与其最接近的是丽水式，同样由珠链串起覆布竹筒和挂有“经幡”的发簪；丽水式与相邻的景宁式和平阳式均有较高的相似度；景宁式将冠顶所覆红布化作錾花银片，竹筒变为竹片，同时将经幡发簪化为主体高冠的一部分；而平阳式除同样将冠顶覆布化作银片外，却同时增大了竹筒的体量，并分别在竹筒前部和竹筒下方增加了银制流苏和红黑条纹头巾；泰顺式继承了平阳式大体量的竹筒主体和额前流苏，同时增加了珠链的用量和长度，在竹筒上方和下方均垂珠链于两侧，且下方的珠链超过1米，长度过腰及臀；福鼎式头饰与泰顺式一样有长长的珠链、额前流苏和脑后飘带，但主体竹筒开始向上倾斜，形成昂起的鸭嘴形；霞浦式头冠倾斜角度进一步加大直至向后，除前额装饰流苏银挂面外，冠顶也装饰银制流苏，松罗头冠则直接利用笋壳做出圆锥形；福安式在霞浦式的基础上继续向后发展，将头冠高度降低的同时纵向长度加长至原来的两倍，后面如霞浦式的冠顶一样垂有银制流苏；同属福安式的宁德头冠冠顶更低，几乎

Based on the former studies and broadened the research of scope and type, we categorized the She costumes in East China and found that the shapes of headwear, earrings, women's shoes and edges of the garment, each has its obvious internal linkage.

Fig. 5-1 shows the development of She headwear. All styles dates back to Luoyuan style, which is similar to Lishui style as they consist of a bamboo crown on the top, bead chains and hairpins in "prayer flag" shape. Lishui style is very similar to Jingning and Pingyang styles. Jingning style replaces the red cloth on the top with a chipped silver and substitutes the bamboo tube with a bamboo blade. That aside, the hairpin is foregrounded as a part of the main body. Pingyang style shares the first alteration while paying more attention to the bamboo by adding more silver tassel to the front, and red and black striped headbands to the bottom of the bamboo tube respectively. Taishun style, similar to Pingyang style, uses a big bamboo tube and keeps the tassel in the front while increasing the length of the bead chain by adding more beads. The bead chain hangs on both sides, above and below the bamboo tube. It spans more than one meter, reaching the waist or even hip. Headdress in Fuding style has a long bead chain, tassel and ribbon, similar to Taishun style. But the bamboo tube tilts upward, looking like an uplifted duck's beak. Xiapu style crown tilts backward while a silver tassel decorates both

呈平顶。到这里，畲族头饰已经由头顶纵向置竹筒再蒙巾的形式，变为平顶阔帽的形式。若从宏观造型上来看，福安凤冠与闽北顺昌的盘瓠帽均呈前高后低的斜平顶状，有一定的相似度。顺昌式、延平式和光泽式均系盘髻于后脑，用黑纱包头，再用红色头绳或带子缠绕，区别是光泽式不用银簪，延平式用6支银钩和1支匙形银簪，而顺昌式用几十甚至上百支匙形银簪。闽南的漳平式受当地汉族影响较大，头饰与客家较为类似。江西樟坪式也相对独特。

畲族耳环造型的变化脉络因其结构简单而颇为清晰（图5-2）。自罗源式开始，到丽水式，一路西进至景宁式，另一路南下到平阳式、福鼎式、霞浦式，其耳坠部分逐渐缩小，形状从葫芦形向箭头形变化，耳后部分变为环状且越来越大。福安式前面的耳坠已近似米粒

the forehead and the top part. Crown in Songluo, which borders Xiapu, is in cone shape by virtue of bamboo shell. Fu'an style further develops based on Xiapu style, reducing the height of the crown while lengthening it to two times of the original one and keeping the silver tassel. Ningde styled crown, belonging to the Fu'an style, has a lower crown, which is almost flat on the top. From here, She headwear has completely changed from the form of a vertical bamboo tube and scarf on the head to the form of a flat one. Generally speaking, the Panhu crown in Shunchang (northern Fujian) and that in Fu'an share similarities as they are high in the front and low in the back. The Shunchang style, Yanping style and Guangze style coil hair in the back of the head, using first black cloth and second (red) hair band. The difference lies in the number of silver hairpins: Guangze style uses none, while Yanping style uses six silver hooks and one spoon-shaped silver hairpin, and Shunchang style uses dozens or even hundreds of spoon-shaped silver hairpins. The style of Zhangping (southern Fujian) is greatly influenced by Han people, and the headdress is similar to that of Hakka. The style of another Zhangping (Jiangxi) is also relatively unique.

As Fig. 5-2 shows, the development pattern of the She earring is simpler than that of the headdress. It, too, originates from Luoyuan and evolves in Lishui. After that, two distinctive categories come into being, with one called

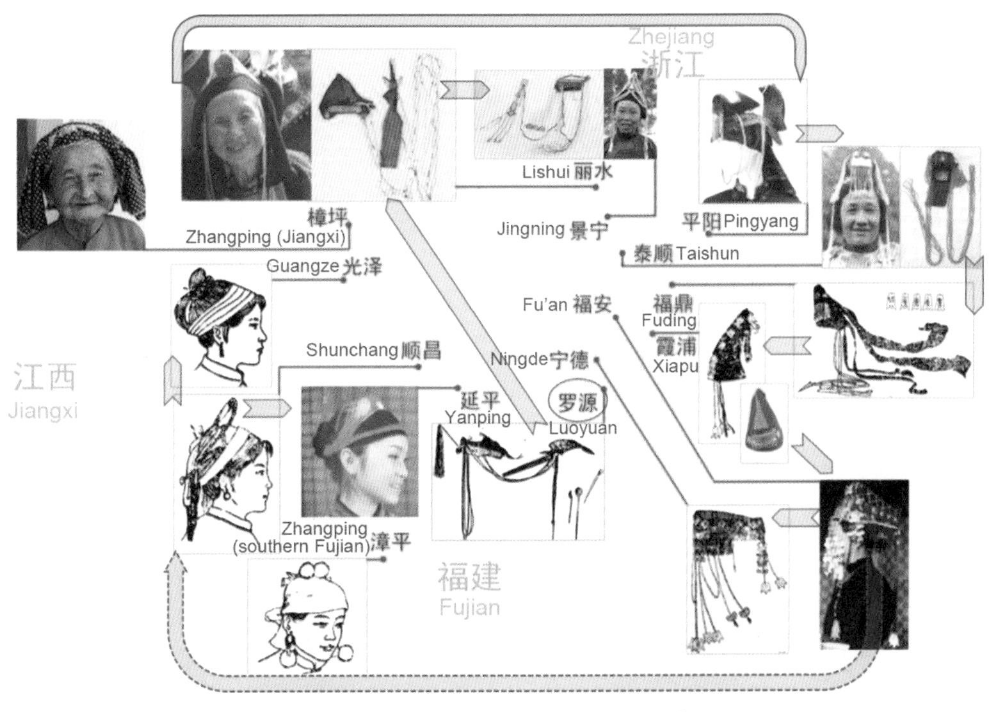

图5-1 华东地区畲族头饰类型分布及变化关系

Fig. 5-1 Distribution and variation of She headwear in east China

大小，而宁德式则环圈更大。发展到顺昌式，耳坠部分已浓缩为环圈前端的一颗小半球。光泽式前端半球较大，后部环圈简化为钩。闽南漳平式耳环与上述各类型联系较弱，自成一脉。

Jingning style and another one represented by Pingyang, Fuding and Xiapu styles. The latter features smaller eardrops and bigger rings in an arrow shape instead of a gourd shape. The eardrops of Fu'an style is as small as grains. Ningde style has bigger rings. When it comes to Shunchang style, the eardrops are rather mini semi-spheres. Guangze style has a pair of relatively bigger eardrops than Shunchang style, but the ring has changed into a hook shape. Earring in Zhangping style (Fujian) differentiates it from others, forming its unique features.

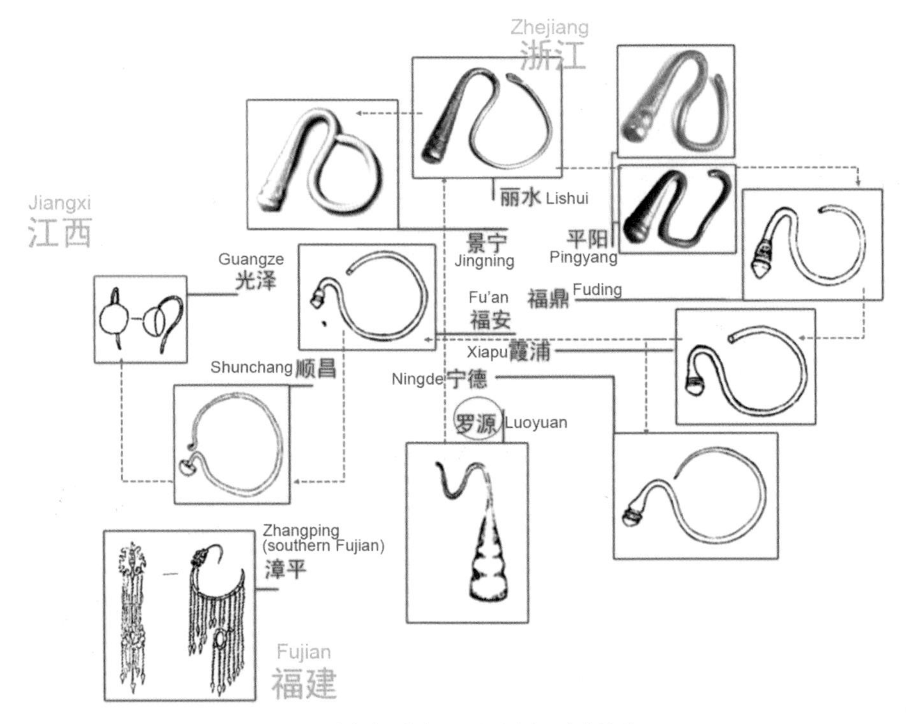

图5-2 华东地区畲族耳环类型分布及变化关系

Fig. 5-2 Distribution and variation of She earring in east China

因资料有限，畲族绣花女鞋造型的变化脉络不够完整，但也大体勾勒出从罗源北上，再从丽水沿东海岸返回闽东的变化路径（图5–3）。罗源式为翘鼻彩须绣花鞋，福安式保留翘鼻而无彩须，丽水式和景宁式保留彩须而无翘鼻，再从平阳式、泰顺式到福鼎式，鞋底前部逐渐上翘，鞋头形状从圆头渐变为方平头。

Based on available sources, it is observed that the development pattern of embroidery women's shoes is not complete. It also starts at Luoyuan and then from Lishui to eastern Fujian along the east coast (Fig. 5-3). Women's shoes in Luoyuan style feature snub noses and colorful fringe laces. The snub nose is followed by Fu'an style while colorful fringe laces by Lishui style and Jingning style. Then comes the Pingyang, Taishun and Fuding styles, whose front part of the sole is gradually warped, and the shape of the toe head gradually changes from a round shape to a square one.

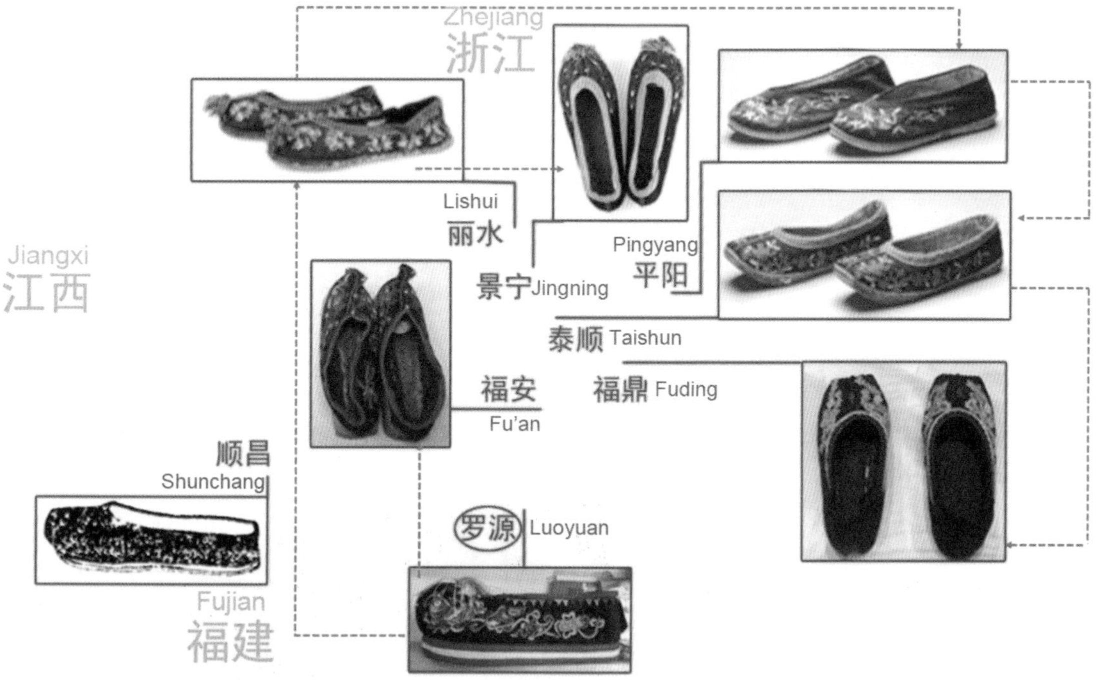

图5-3 华东地区畲族女鞋类型分布及变化关系

Fig. 5-3 Distribution and variation of She women shoes in east China

如图5-4所示，虽然女装前襟造型受主流服饰文化涵化程度颇大，导致脉络不清晰，但仍可见其地方性差异的内在联系：从浙南丽水、景宁，下至闽东福安，再到闽北顺昌、延平，最后至闽南漳平，沿线及向西各地畲服均不同程度地涵化为近代主流的旗式右衽系扣大襟衣，前襟呈“厂”字形，并沿襟边贴边或绲边装饰。而该沿线以东地区畲服明显呈现出畲族自身特色与主流服饰特征的融合：罗源式保留汉式直襟交领造型，而脖颈周围挖出仿旗式

As shown in Fig. 5-4, although the front of women's garments is greatly influenced by mainstream clothing culture, leading to not so clear development pattern, the internal connection of local differences can still be seen. From Jingning, Lishui in southern Zhejiang to Fu'an in eastern Fujian, Shunchang, Yanping in northern Fujian, and Zhangping in southern Fujian, places along the route and westwards experience varied acculturation and embrace wide garments with right lapel, featuring L-shaped front decorated by piping. The eastwards of the route exhibit the

圆领结构；霞浦式、福鼎式、平阳式和泰顺式前襟均不用纽扣，为汉式腋下系带形式，但直襟演变为钝角，呈现出向旗式大襟的过渡。

integration of local She costumes and mainstream clothing culture. Luoyuan style keeps certain characteristics of Hanfu: stand-up collar with a straight edge, but develops the collar into a round one, or mandarin collar in cheongsam; Xiapu style, Fuding style, Pingyang style and Taishun type uses no buttons in the front but lace up in the armpit. The shape of the front has changed from a straight edge to an obtuse one, a transition to the front of the cheongsam.

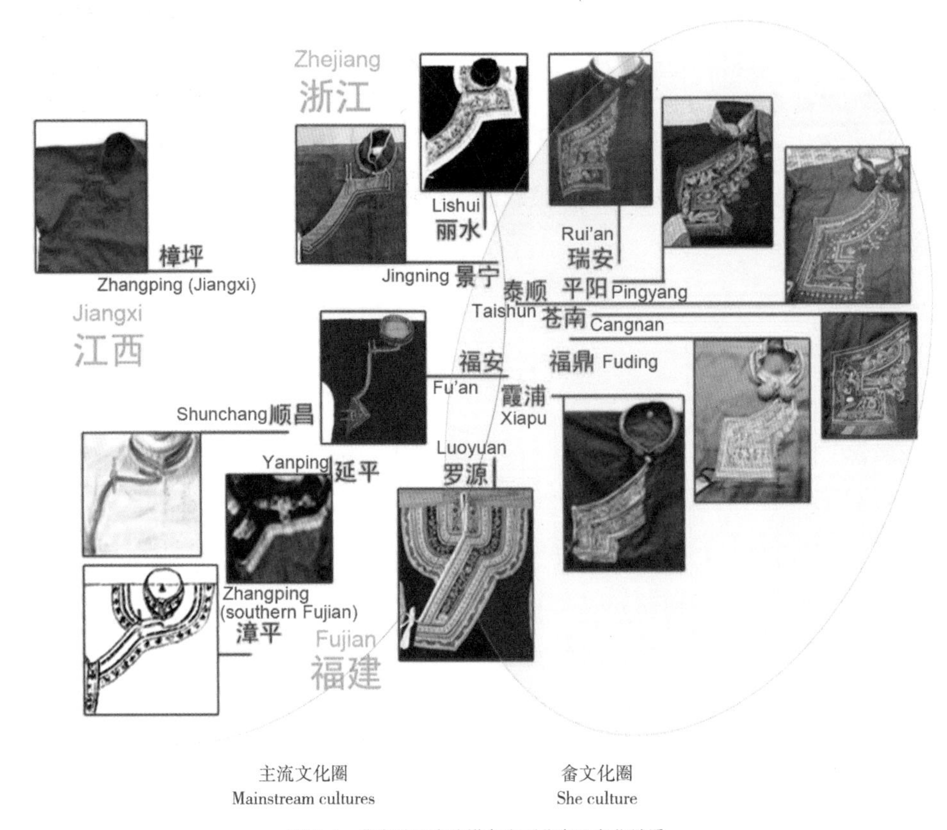

图5-4 华东地区畲族襟角类型分布及变化关系

Fig. 5-4 Distribution and variation of the front of She costumes in east China

综上所述，华东地区畲族各类服饰造型的变化脉络大体上呈现出以罗源式为源头，直抵丽水后沿东海岸返回闽东，再由福安向西辐射的变化路径。这说明了畲族在历史上的迁徙和各族间影响并不仅仅是递进式的，而极有可能是阶梯式和回溯式的。同时，通过对比华东地区畲族各类服饰的造型，也可发现主流文化圈和畲族文化圈的边界大体上位于从浙南丽水、景宁，下至闽东福安，再到闽北顺昌、延平，最后至闽南漳平的沿线上。

Based on the analysis above, it can be summed up that the development pattern of She costumes in east China basically originates from Luoyuan, and evolves in Lishui, then back to eastern Fujian along the east coast, and then stretches westwards from Fu'an. This shows that the migration and inter-ethnic influence of She people in history is not only progressive, but probably stepped and retrospective. That said, by comparing the structure of various She costumes in east China, it can also be found that the boundary between mainstream culture and She culture borders on the line from Jingning, Lishui in southern Zhejiang, to Fu'an in eastern Fujian, Shunchang and Yanping in northern Fujian, and finally Zhangping in southern Fujian.

练习题 | Exercise

请将以下畲族服饰与其穿着地域用线连起来。

Please match the She costumes with their birthplaces.

浙江景宁 Jingning, Zhejiang	浙江丽水 Lishui, Zhejiang	江西樟坪 Zhangping, Jiangxi	浙江麻江 Majiang, Zhejiang

福建罗源 Luoyuan, Fujian	福建福鼎 Fuding, Fujian	福建顺昌 Shunchang, Fujian	福建福安 Fu'an, Fujian	福建霞浦 Xiapu, Fujian

参考文献

References

[1] 雷志良．畲族服饰的特点及其内涵［J］．中南民族学院学报（人文社会科学版），1996（5）：131.

[2] 沈其新．图腾文化故事百则［M］．长沙：湖南出版社，1991.

[3] 360百科．畲族凤凰装·历史渊源［EB/OL］．2020-09-28.

[4] 凌纯声．畲民图腾文化的研究［C］//中国边疆民族与环太平洋文化．台北：联经出版社，1979.

[5] 潘宏立．福建畲族服饰研究（油印本）［D］．厦门：厦门大学，1985.

[6]《浙江省少数民族志》编撰委员会．浙江省少数民族志［M］．北京：方志出版社，1999.

[1] Lei Zhiliang. The characteristics and connotation of She costumes[J]. Journal of South-central University For Nationalities (Philosophy and Social Science), 1996(5): 131.

[2] Shen Qixin. One Hundred Stories of Totem Culture[M]. Changsha: Hunan Publishing House, 1991.

[3] 360 Baike. History of She Phoenix Costumes [EB/OL]. 2020-09-28.

[4] Ling Chunsheng. She Totem Culture Study [C]// Chinese Frontier Nationalities and Pacific Rim Culture. Taipei: Lianjing Publishing House, 1979.

[5] Pan Hongli. Research on Fujian She Clothing (mimeographed)[D]. Xiamen: Xiamen University, 1985.

[6] Zhejiang Provincial Ethnic group Records Compiling Committee. The Records of Ethnic Minorities in Zhejiang Province[M]. Beijing: Publishing House of Local Records, 1999.

[7] 钟雷兴，吴景华. 闽东畲族文化全书：服饰卷[M]. 北京：民族出版社，2009.

[8] 伊莎贝拉. 长江流域及其腹地[M]. 剑桥：剑桥大学出版社，2010.

[9] 郭柏苍. 闽产录异[M]. 长沙：岳麓书社，1986.

[10] 范佩玲. 山哈风韵——浙江畲族文物特展[M]. 北京：中国书店出版社，2012.

[11] 陈敬玉. 民族迁徙影响下的浙闽畲族服饰传承脉络[J]. 纺织学报，2017，38(4)：115.

[7] Zhong Leixing, Wu Jinghua. Complete Book of She Culture in Eastern Fujian: Costume[M]. Beijing: The Ethnic Publishing House, 2009.

[8] Isabella L. The Yangtze Valley and Beyond[M]. Cambridge: Cambridge University Press, 2010.

[9] (Qing Dynasty) Guo Baicang. Records of Fujian Production[M]. Changsha: Yuelu Press, 1986.

[10] Fan Peiling (Ed.). Shanha Chance: Zhejiang She Cultural Relics Special Exhibition[M]. Beijing: China Bookstore, 2012.

[11] CHEN Jingyu. She's costumes heritage context in Zhejiang and Fujian provinces based on ethnic migration history[J]. Journal of Textile Research, 2017, 38(4): 115.